Step Up Your Teamwork

STEP UP
YOUR
TEAMWORK

FRANK VISCUSO

Fire Engineering

Disclaimer

The recommendations, advice, descriptions, and the methods in this book are presented solely for educational purposes. The author and publisher assume no liability whatsoever for any loss or damage that results from the use of any of the material in this book. Use of the material in this book is solely at the risk of the user.

Copyright ©2015 by
PennWell Corporation
1421 South Sheridan Road
Tulsa, Oklahoma 74112-6600 USA

800.752.9764
+1.918.831.9421
sales@pennwell.com
www.FireEngineeringBooks.com
www.pennwellbooks.com
www.pennwell.com

Marketing Manager: Sarah De Vos

Director: Mary McGee
Managing Editor: Marla Patterson
Production Manager: Sheila Brock
Production Editor: Tony Quinn
Cover Designer: Charles Thomas

Library of Congress Cataloging-in-Publication Data

Viscuso, Frank.
 Step up your teamwork / Frank Viscuso.
   pages cm
 Includes index.
 ISBN 978-1-59370-354-7
 1. Fire departments--United States. 2. Fire departments--United States--Management. 3. Fire departments. 4. Fire departments--Management.
 5. Teams in the workplace. I. Title.
  TH9503.V57 2015
  363.37068'4--dc23
                                        2014040266

All rights reserved. No part of this book may be reproduced, stored in a retrieval system, or transcribed in any form or by any means, electronic or mechanical, including photocopying and recording, without the prior written permission of the publisher.

Printed in the United States of America

3 4 5 6 7 8   21 20 19 18 17

"Quit! Give up. You can't do it. You're too slow, too fat, too ugly, too stupid, too lazy, and too late. You're unrealistic!"

Do you know that little voice, the one inside your head that tells you that you can't achieve your goal? This book is dedicated to those brave people who are committed to defeating that voice. When you conquer fear, self-doubt, and self-induced negativity, you will master your life. When you surround yourself with others who have done the same, you will have a team that is capable of achieving anything.

Contents

Foreword by Anthony Kastros xi

Acknowledgments xiii

Introduction 1

1 Teamwork in the Fire Service 7
Mission First 12
Review, Evaluate, and Revise 21
You Can't Achieve Significance without Sacrifice 22
Define Your Team: The Rule of Five 24
Life Enhancers vs. Lawn Mowers 26
Communication, Coordination, and Control 28
What Stage Is Your Team In? 34
Teamwork Strategies 38
Creating Momentum 41
Team-Building Exercises 46

2 Preparing for Success 63
Decisions Are Like Tattoos 64
Activity vs. Productivity 66
After Action Review 70
Root Cause Analysis 72
How We Learn 74
Preparation Leads to Confidence 87
Sweat More, Bleed Less 88
Daily Method of Operation 90
Eat the Ugly Frog First 92
Never Settle 93
The Best Ideas Have to Win 95
What's Your Brand? 97
Marketing Your Company 99
Ten Qualities of an Outstanding Team Player 101
Mentorship Programs and Succession Planning 106
An Uneducated Firefighter Is a Dangerous Firefighter 110

3 Leading Teams ... 113
What's Your Title? ... 113
Leadership 101 ... 116
Are You Ready for a Leadership Role? ... 117
Ten Signs a Person May Not Be Ready to Lead ... 118
The First Follower ... 123
Change Is Necessary ... 125
Change Can Be Good ... 127
How to Lead an Organization Through Change ... 129
Transparency ... 132
Transparency in the Digital Age ... 139
Specific Intent ... 139
Understanding People ... 140
The Lost Art of Listening ... 149
Bonding ... 150
Give More Responsibility ... 153
Establishing Team Expectations ... 154
Priorities ... 158
The Headline Test ... 159
The Power of Words ... 160
Communication ... 163
The Power of Stories ... 165
Public Speaking ... 169
The Attention Span of a Goldfish ... 170
High Tech vs. High Touch ... 171

4 Preventing Team Collapse ... 173
"Routine Fires" Don't Exist ... 174
Why Teams Fail ... 179
Dysfunctional Teams ... 186
Team BLEVE ... 188
Conflict Resolution ... 189
Handling Difficult People ... 196
Accountability ... 207
Climate ... 219
Dialogue, Discussion, and Debate ... 220
Dumb Things Firefighters Say ... 221
Multiple Alarms—Call for Help Early ... 222

5 Building Your Legacy ... 225

- Turn Adversity into Advantage ... 226
- Failure Isn't Fatal ... 229
- Courage under Fire ... 233
- Combustible Courage ... 235
- What's Your One Degree of Difference? ... 237
- It's Not Over until We Win ... 240
- Commitment ... 242
- Service to Many Leads to Greatness ... 244
- Your Needs Come Last ... 248
- A Negative Attitude Is Poisonous ... 250
- Right Is Right ... 252
- DTRT ... 254
- Pushing vs. Pulling ... 255
- A Culture of Hatred ... 257
- The "I" in Team ... 260
- Be a Culture Creator ... 262
- The Step-Up Challenge ... 266

Index ... 273

Foreword

A loss of leadership is sweeping the American fire service. Mass attrition of experienced members, a national recession, task obsession, and lack of adequate succession planning are just some of the contributing factors to this perfect storm.

Occasionally, tragedies occur like sudden torrential downpours. All too often we are notified of a disaster in which multiple firefighters have died, and along the way, some have become indelibly etched into our souls. There are cities whose names alone send shivers up the spine, including Houston, Charleston, Toledo, New York, San Francisco, Prescott, and countless more.

Many mistakenly believe that tactical operations and leadership are two distinctly separate, stand-alone components of this job. Nothing could be further from the truth. Effective fireground operations are built upon strong leadership, long before the bell rings. Excellent leaders who build excellent teams perform with excellence on the fireground. It's that simple.

Fire service leadership is exponentially more challenging than ever before. Unprecedented public scrutiny, technological permutations of all kinds, and lightning-fast speed of information all add to the challenges faced by fire service leaders. To compound the problems, we have failed to prepare our leaders for the predictable challenges that they will face, let alone the unpredictable. New officers are making the same old mistakes, while being challenged with problems that are new to us all.

Enter Chief Frank Viscuso, who truly is a renaissance man. He has the mind of a poet, the heart of a lion, and the passion of Leonidas. He has a God-given and uncommon blend of humility, street credibility, experience, education, and above all, passion.

Frank has dedicated his life to the art of fire service leadership, no matter the form. In his best-selling book *Step Up and Lead*, he gave the new millennial generation and old salts alike a treasure trove of tools to handle practical, real-world problems that plague our beloved service. God's gifts to Frank have become blessings to the fire service.

If you visit his Facebook page or attend one of his workshops, you encounter a man who has seen a national problem and chosen to grab it by the throat. He tirelessly brings words of inspiration, insight, candor, humor, and wisdom from countless experiences and sources.

In *Step Up Your Teamwork*, this wonderful complement to his previous work, Chief Viscuso has delved further into the heart of real-world fire service leadership. Frank knows that the dimension of teamwork is so vast and vital to accomplishing the mission that he has dedicated a year of his life to addressing this topic in detail. Like a thermal imaging camera, this book sees through the fog and brings into view the unique spectrum of light that is team excellence.

Without vibrant, aggressive, passionate teams, the mission will fail. When the mission fails, Americans die. Whether within the military or our sacred calling, the fire service, the mission can only be successfully accomplished if leaders know how to build and lead teams.

This is not a hobby or sport where everyone gets a ribbon. This is not a job where you can regurgitate some tired old jargon, or parrot sophomoric slogans like "Leader's Intent" or "Lead by Example" or "Be Accountable" and then leave the room. You must define leadership, embody it, and show your team how to accomplish it. Otherwise, it's just ear candy.

Frank Viscuso explains in careful detail how to build, lead, repair, rejuvenate, empower, and even savor the most critical component of the fire service: the team.

As you will see in the pages that follow, Frank uses his vast array of experiences to tell stories that draw you into the moment and make the point crystal clear. His unique blend of fire service street credibility and private sector experience has resulted in wisdom that we may draw from in our pursuit of leadership excellence.

Thank you, Chief Viscuso. You are filling the tactical gap between the strategic and task levels, where vision transforms into reality, where theory becomes practical, and most importantly, where lives will be saved.

God bless you and your precious family!

Anthony Kastros

Acknowledgments

Throughout my life I have been lucky enough to belong to some great teams. One of them is the team at PennWell Books, specifically Mary McGee, Marla Patterson, Cindy Huse, Tony Quinn, and my brother from another mother, Mark Haugh. It has been a tremendous honor to be a small part of the best team in the publishing industry. Thank you for all your hard work and dedication.

Another great team I have been blessed to be a part of is the Kearny Fire Department, specifically Group C, which is built upon a foundation of courtesy, courage, and commitment. You are some of the hardest working, most creative, and fun-natured guys I have ever known. It's been an absolute pleasure working alongside you.

My top team is the one I think about every moment of every day of my life—my family. To my wife Laura and our boys Thomas, Nicholas, and Frank Jr., you are my strength and motivation. Thank you for your unwavering, unconditional support. We've had our highs and our lows, our victories and our challenges, and we have always found a way to get through it all, together. It's been an awesome ride so far, and we have so much to look forward to.

Most of what I learned about team building came from my mother and father. I am one of four children. Through their actions, my parents taught us about staying together, celebrating our victories, rewarding good performance, having the right attitude, balancing work and play, encouraging each other, delegating tasks based on our strengths, being committed to each other, setting goals, persisting, and practicing delayed gratification in a society that demands satisfaction in an instant. My parents know a little something about strong leadership, but what makes them different from most is they also know that success at the expense of others is not success. The only way to achieve true success in life is to be part of a team where everybody wins. Thank you, Mom and Dad. I am forever grateful for the life lessons you have taught me. Now it's time for me to pay it forward.

INTRODUCTION

We all know that building a championship-caliber team is the quickest way to reach our goals. Why then, do so many people in leadership positions fail when it comes to team building? Failure is not necessarily due to a lack of desire or effort on the part of leaders; on the contrary, they may have put a great deal of time and sweat equity into their cause. The truth is most people simply do not know how to develop and motivate a team. Others who do understand team building are not always able to maximize the results that can be achieved when a group of people come together to work toward a common goal.

Teamwork is essential for the survival of an organization. This is as true in the workplace as it is at home or on the playing field, and it certainly holds true in the fire service. When people think about firefighters, they also think about teamwork and bravery. The way a team of firefighters responds to a fully involved structure fire, develops a strategy, divides into smaller teams, and tackles a variety of essential tasks to accomplish their goals exhibits this solidarity and courage.

In my opinion (and based on personal experience), the fire service has produced some of the most efficient teams in the country, but establishing the highest performance level of a team does not come easily, even in our industry. Most organizations within corporate America come together for 6 to 8 hours a day, work on a project, and return home. Firefighters can spend up to 48 straight hours with each other. They train together, eat together, run community events and fund-raisers together, develop policies and procedures together, fight fires together, respond to medical emergencies together, and solve problems together, 24/7/365. In fact, every 15 seconds firefighters respond to a working structure fire somewhere in the United States.

Firefighters in general are resilient people with great determination and the ability to thrive and succeed in the most adverse conditions. They believe in a code of honor called *brotherhood* (sisters included). The Oxford Dictionaries online defines *brotherhood* as "the feeling of kinship with and closeness to a group of people or all people; 'a gesture of solidarity and association.'" For us, it is more; it is an unbreakable bond that is stronger than most civilians could possibly imagine. There is an absolute reason why this connection between firefighters is so strong. We understand that at any given moment, we may be expected to risk our lives for strangers or for each other. This is something that all firefighters learn during the first week of their academy training and hopefully never forget.

As with most of corporate America, major changes continually challenge the fire service, including downsizing, enormous competitive demands, extraordinary pressure, technology advancement, and even struggles to stay in business altogether. The fire service has never been stagnant or content. We are constantly changing and adapting. Firefighting teams are famous for adapting to whatever situation they respond to. Their job is to figure out the best way to accomplish their goal with the resources available and do it as fast as possible. They take calculated chances based on a risk vs. reward analysis, and they allow very few obstacles to be utilized as acceptable excuses for failure.

I have the privilege to serve as a deputy chief and tour commander in the fire service in one of the busiest and most densely populated counties in New Jersey. As such, my job is the same as that of every other tour commander in the country. We pull together a team from individuals ranging from overachievers who feel firefighting is their calling to underachievers who prefer to do the bare minimum on any given day. To be an effective team leader, I consider it my job to embrace the awesome power of teamwork. I accomplish this by identifying and encouraging each person to tap into their individual strengths and teaching them how to utilize their talents, skills, and abilities for the betterment of the team. If you are reading this book, you may share a similar type of responsibility.

Introduction

Leading a team can be an overwhelming task for some. Early on, it was difficult for me, but through self-education, practice, and trial and error (both in and out of the fire service), I have discovered what works and what never will work when it comes to building a successful team.

The citizens we protect do not expect and will not accept a mediocre response when they call the fire department, and neither should the men and women who wear the uniform. People expect us to perform like a championship team every time we show up. Think about what would happen if we arrived on scene, looked at each other with puzzled expressions, and then shrugged our shoulders because we did not know what to do. Everyone around us would panic. Part of our job is to remain cool, calm, and collected—courageous under fire. It is a fire service leader's job to prepare his or her team to give "more than expected" every time the alarm sounds. Our customers expect the highest level of service, and the goal of this book is to give you great insight as to how we not only meet but exceed their expectations. I want to teach you how you can bring your team to the highest possible level and harness *the power of one degree*.

One Degree of Difference

In the science of fire behavior, the number 212 is pivotal; it represents the difference between ordinary and extraordinary. Firefighters immediately relate this number to a boiling pot of water. Imagine that the same boiling water represents a high-functioning team. Remove one degree and that same pot of water is just hot water. Most teams fall just short of their potential, when one degree more may have made all the difference.

While the boiling point of water is defined to be 212°F, consider for a moment that most common flammable and combustible liquids have autoignition temperatures. The autoignition temperature or kindling point of a substance is the lowest temperature at which it will spontaneously ignite in a normal atmosphere without an external source of ignition, such as a flame or spark.

The autoignition temperature of ethanol is 685°F, gasoline is 536°F, diesel is 410°F, jet fuel is 410°F, and kerosene is 428°F. As with the difference between 211°F and 212°F in terms of boiling water, a difference of just 1°F is all it takes for full autoignition of the liquids listed above. If 1°F is all it takes to make such a significant difference in these examples, the same could be said for success in team building.

Consider for a moment that the difference between success and failure often comes down to one degree more of effort. One extra degree of effort often separates the good from the great. Consider the following examples of how one extra degree made all the difference for both teams and individuals.

According to Biography.com, "Swimmer Michael Phelps has set the record for winning the most medals, 22, of any Olympic athlete in history. . . . He went on to win medals at the Olympic Summer Games in Athens, Beijing, London, and Rio de Janeiro, accumulating a total of 28 medals and setting the record for the most medal wins by any Olympic athlete. . . . Phelps also holds the record for the most gold medals won in a single Olympics (eight gold medals at Beijing in 2008)." Of course there were races where he dominated his competition throughout his storied career, but most of the time his margin of victory was less than 1 second. In fact, when Phelps won gold in the 100-meter butterfly in 2008, it was by 0.01 second over Milorad Čavić of Serbia—one degree of difference.

The top golfers in the world are often only one stroke ahead of the pack, but can earn millions more than the others. The average margin of victory at the Indy 500 is less than 2 seconds. In the 1992 Indy 500 race, Al Unser Jr. won over Scott Goodyear by only 0.043 second.

In the 1984 Olympics, Michael Gross, nicknamed "The Albatross" for his 6'7" frame, was part of West Germany's highly publicized 4 × 200-meter freestyle relay team. Anchored by Bruce Hayes, the Americans edged out Gross and the West Germans by only 0.04 second.

According to Academic Dictionaries and Encyclopedias online (http://en.academic.ru/), in 2005, "Paul Tergat of Kenya barely outsprinted Hendrick Ramaala of South Africa in the final meters of the New York City Marathon for a time of 2:09:30, beating Ramaala by 1 second."

I have personally competed in collegiate rowing competitions where the difference between 1st and 4th place was less than 1 second. All are examples of one degree of difference.

When you stop to think about it, even great individual accomplishments in life are not achieved without the help of others. Phelps, for example, had teammates and coaches. In fact, many of his medals were won as a member of a relay team. Behind every winner is a team. A collegiate running back may have won the Heisman trophy, or an NFL quarterback may have been chosen as the Super Bowl MVP, but both had coaches and other players on the field with them. Alongside every great achiever, you will find people who have played a role in that individual's success.

Teams are everywhere. You will find them in homes, businesses, and stadiums. They exist to serve a purpose, and they either fail or succeed together. Just showing up and punching a time clock does not make someone part of a team. Neither does the simple act of putting on a firefighter's uniform. You and your team must be unified by a common goal. Showing up is not enough. Without shared values and common goals, peak performance is not possible. What would happen if you and everyone on your team put in one extra degree of effort? How efficient could your team become? Is it even possible that a slight change in philosophy can make a huge difference?

I believe the answer is yes. Let's explore it together as we discuss various proven team-building methods you can begin to implement immediately. Throughout this book, I want you to constantly remind yourself that the margin of victory is often so slim that one degree can make all the difference. Ask yourself, "What is that one extra degree of effort that could change my team's productivity level from hot water to steam?"

TEAMWORK IN THE FIRE SERVICE

1

You must learn how to hold a team together.
You must lift some men up, calm others
down, until finally they've got one heartbeat.
Then you've got yourself a team.

—Paul William Bryant

It was a quiet night in the firehouse, and quiet nights are uneasy. It may seem odd to link those two things together, *quiet night* and *uneasy*, but firefighters understand the connection. You should know before I share this story that just about every multiple alarm fire or intense rescue story begins this way. It is so common that if a probationary firefighter makes reference to how quiet it is around the firehouse, the experienced firefighters will roll their eyes in disgust and reply, "Oh great, you just jinxed us!"

It was after 1:00 AM when three separate alarms came in simultaneously. The first call was a report of a downed power line. The second was a water flow alarm in a warehouse, and the third was a report of a man whose leg was pinned between a loading dock wall and a semi-trailer. The tour commander was automatically dispatched by the computer-aided system to the water flow alarm. However, water flow alarms in that warehouse were almost nightly occurrences. A man pinned to a wall by a trailer was not, so the tour commander made the obvious decision to respond to that call instead, along with an engine and ladder company consisting of a total of five members.

The tour commander arrived on scene about one minute before the other companies and quickly assessed the situation. A heavyset man was working on the loading dock. He was carrying boxes from the warehouse into a trailer that was backed up against the building. He tried to step on an extender that bridged the gap between the dock and trailer. Instead, he stepped between the extender and the platform stop, and his leg fell between the trailer and the platform. His body dropped down in an awkward position, and his upper thigh became trapped. Every time he moved, he slid further down, causing the platform to clamp his leg more tightly (fig. 1–1).

Fig. 1–1. This is a space similar to the one where the worker's leg was trapped.

It would have been easy to pull the trailer away from the building and free the man's leg. The problem was that there was no truck attached to it. If the firefighters tried to make the connection between the two now, it might cause the trailer to move, which could very well end up severing the man's leg.

The man was in agony. He had been there for nearly 30 minutes before we received the call. The police and EMS had been standing by as coworkers tried to free him, but every time they tried, they ended up making it worse by tightening the gap.

"Get me out of here!" he cried. "I need to get out of here, now!"

"Be patient for a few more minutes," the tour commander said. "We will get you out as soon as our equipment arrives."

As he finished the sentence, two companies pulled onto the property, and the captain walked up to the tour commander to discuss the situation. In the corporate world, this captain would be the type of manager everyone would want to have on staff. He was educated, reliable, proactive, ambitious, and willing to get his hands dirty 24/7. He was a lead-by-example type of guy.

At this particular incident, knowing a man's limb was entrapped, the tour commander was happy that this captain was one of the members on duty. He was well versed on extrication and rescue. They met like they always do, for a face-to-face consultation to look at the situation and determine their best course of action together. They both agreed that the "airbags" would be their best option. The airbag system is an air-operated lifting tool used to assist in the physical rescue of victims in a variety of situations. They are extremely strong and durable bags that have the power to lift, move, or shift weights up to 70 tons.

Three team members gathered the equipment. Another one cleared the area of any unnecessary people, while the captain notified the other emergency service personnel of their plan. They lifted the platform and located a solid position on the front of the loading dock. The firefighters quickly but carefully placed the bags against the building opposite the trailer. It was time to put the plan into effect.

While they were preparing the operation, the tour commander leaned down to make eye contact with the victim and explained what actions they were going to take. He also took the initiative to put a plan B into effect. If the airbags did not work, for one reason or another, the only other option would be to pull the trailer away from the structure with a heavy-duty tow truck. Since the company did not have one on site, the tour commander called for one from a local towing company. It was early enough in the morning that they would have to call someone into work, get the truck, and respond to the scene. This would take 30 minutes or more, and the popular practice in the fire service is to call for resources early. It is better to have more resources than you need than it is to lack a critical resource. They could always turn them back if they did not need them.

Even though he set plan B in motion, the tour commander was confident in plan A. They all were. Well, the firefighters were, but the same could not be said for one of the warehouse employees. While the tour commander was assuring the entrapped man that we were going to have him free within a few minutes, the firefighters encountered their first major obstacle—a warehouse manager who took it upon himself to tell them, and the entrapped man, that their plan would not work.

"You're going to cut his leg off if you try that!" the manager said, loudly enough that everyone in the area could hear him. Then he turned and walked away, repeating his doubt-filled comment several times from a distance.

Of course, this was the last thing the trapped man needed to hear. The tour commander again assured him that the plan was going to work, but the words of his colleague clearly frightened him. The manager's words, however, did not affect the firefighters. When a team is well prepared, and they believe in each other's judgment and abilities, there is no room for doubt, especially when it is time to execute the plan. They tuned the manager out and went to work.

They set up their tools, positioned the airbag, and when the captain gave the word, another firefighter filled the bag with air. Within seconds, the man's leg was free. When the tour commander walked into the main office a minute later to get information for our report, their critic again said, "Your plan isn't going to work, you're going to cut off his leg."

The tour commander pointed out the office window at a large man who was gingerly walking toward the ambulance and replied, "Whose leg, that man's?"

The look on the manager's face was priceless.

I love it when the members of a team believe in themselves and the team successfully executes a plan in spite of what people around them are saying (fig. 1–2). For many years, I have shared an inspirational quote with teams and individuals that sums up my feelings about negative people like the manager the firefighters encountered at this incident. The quote is, "People who say it cannot be done should never interrupt people who are doing it." This manager was a problem finder, and the first rule of developing a successful team is to surround yourself with problem solvers.

Fig. 1–2. Belief in your ability to execute effectively as a team in stressful environments is only attainable with practice and experience. *Courtesy:* Ron Jeffers.

On a side note, two of the five firefighters working that day were probationary firefighters who had only been on the job for a few months. They operated in an extremely confident and professional manner, which was a result of the aggressive training schedule they had been on that required them to participate in a number drills with their rescue equipment.

So to sum it up, the team was successful because we followed this simple six-step process:

1. Assess the situation.
2. Establish your goal.
3. Develop a plan of action.
4. Believe in each other, your tools, and your training.
5. Execute your plan without hesitation.
6. Review, evaluate, and revise when necessary.

These six areas will be covered throughout this book. As you continue reading, I will share some intense fire stories. The one you have just read is not very intense, but it illustrates the point of how we as firefighters work together at an incident and tune out negative noise to accomplish our goal.

People who say it cannot be done should never interrupt people who are doing it.

—*Unknown*

Mission First

The cornerstone of the fire service is teamwork, and the key to teamwork in the fire service is to put the mission first. This must be evident in every aspect of our profession, from training exercises to multiple alarm fires. The mission is why we exist, and the mission changes from call to call, which is why we must constantly train. I believe the fire service is the ultimate definition of the word *teamwork*. To understand why I believe this, a person would have to experience or witness what happens from the time of the initial call until all companies return to the station.

The fire service is a large team composed of many subteams (fig. 1–3). Here is a scenario to illustrate the how many moving parts have to come together to achieve success on the fire ground. A 9-1-1 dispatcher receives a call that a house is on fire and begins to gather as much information as possible. She begins to assess the situation. Who is making the call—a resident, a civilian passing by, or a police officer? Is anyone in the structure? Is flame or smoke visible? Are all the residents accounted for?

Fig. 1–3. A fire department is a large team composed of many subteams. Various moving parts have to come together in order to achieve success.

Once information is gathered, the operator will set the ball in motion by dispatching specific companies to the scene. The number and types of companies vary from department to department. One organization may dispatch six firefighters on two apparatus. Another may send two dozen firefighters on six apparatus. It all depends on the size of the fire, the size of the department, and the availability of firefighters and resources. Whatever is sent, it will be (at a minimum) a combination of engine and ladder companies, as well as at least one officer who will act as the overall incident commander (IC).

These companies, often housed in different fire stations, train together on a regular basis so when they are called to act they can seamlessly come together and work toward a common goal. Firefighters also spend many hours preplanning buildings and areas and assembling information that will help them determine their strategy months, even years, before the alarm comes in. This will include information such as hydrant locations, nearby exposures, life hazards, and also information regarding sprinkler, standpipe, and utility control (fig. 1–4). On the way to the fire scene, all firefighters will monitor their radios so everyone will be aware of what is happening around the incident. The more information the team can acquire before stepping out of the apparatus, the better off they will be.

Fig. 1–4. Preplanning is a major component to preparing a team for success. *Courtesy:* Lars Ågerstrand.

The first officer to arrive will establish command, and he or she becomes the incident commander. The IC is equivalent to an organizational leader. If this were a football game, the IC would be the coach. This is the person who identifies the needs, determines the strategy, and calls the plays. A fire incident needs an IC, but the person who establishes command is not necessarily the one who will remain in command. Once a higher ranking officer arrives on scene, that person often assumes the responsibilities of the IC and becomes the new team leader. The first-arriving IC will give an initial radio report to inform the dispatcher and the other incoming units about the situation. This report will include number of floors, construction type, visible smoke or fire conditions, whether the building is occupied or unoccupied, and what strategy they are going to implement. This is also the time to call for any additional resources that may be needed.

Up to this point, a lot has happened, but this is just the beginning of a long incident that will require intense dedication and a tremendous display of teamwork. As engine and ladder companies arrive on scene, they will be assigned different tasks. Before those assignments are given, the IC must complete a thorough assessment to ensure that every initial concern is identified and addressed. The 15 points of consideration are the following: construction type, occupancy (contents), apparatus and staffing needs, life hazards, terrain, water supply, auxiliary appliances, street conditions, weather conditions, threatened exposures, area and layout of the building, location and extent of fire, time of day, height of building, and any special considerations.

There is a lot to consider, and the IC must be dedicated to self-education so he or she is fully capable of properly assessing the situation and requesting the necessary resources. These resources may include, but certainly will not be limited to, the following:

- Second or third alarm, to bring more firefighters to the scene
- Mutual aid companies to cover the town, while they fight the fire
- Utility company personnel to shut down or control the gas, electric, and/or water supply to the building
- Law enforcement personnel for traffic or pedestrian control
- EMS for patient triage, treatment, and transportation
- A rapid intervention crew to rescue firefighters who may become trapped or distressed
- A safety officer to act as additional eyes and ears to help the IC recognize any unsafe acts or conditions
- A water supply officer if the fire is big enough to need multiple water supplies
- An accountability officer to help track and account for all fire and emergency service personnel working in and around the fire building

- A rehabilitation officer (and unit) to provide an area for rest, food, and replenishing fluids

Those are the most common resources called to the scene, but there are many others that may be needed, such as the Red Cross to help displaced occupants, a hazardous materials team to assist with chemicals or other toxic substances discovered, an urban search and rescue team if the building collapses, among others.

While all of this is happening, the IC (team leader) is giving out assignments to the other firefighters and companies on the fireground. There are a number of important tasks that need to be accomplished in order for this team of firefighters to achieve their overall goals, which could be summed up in three categories: life safety, incident stabilization, and property conservation (fig. 1–5).

Fig. 1–5. The incident commander gives each company an assignment and then communicates with a number of people to ensure the incident goals and objectives are being accomplished. *Courtesy:* Ron Jeffers.

To accomplish those goals, every team member and every engine and ladder company must do their part. Before continuing our discussion, it may be helpful to briefly define a couple of terms.

The first term often used when discussing leadership is the word *manager*. In my opinion, the word should be changed to *team leader*. When I speak to corporations, I often say that once a person moves from productive mode to management mode, the game is over. In today's society, too many people mistake the word *manage* for *maintain*. It is more than that. A true leader in a managerial position motivates, sets goals, and directs or redirects the vision for the team. Those qualities go beyond what most Americans think of when they hear the word *manager*. I know we need to keep the word as a title for pay reasons, business cards, and corporate structure, but we do not have to use it around the workplace when we are talking about team building.

The second term we should discuss is *layered leadership*. It is not enough for a team to have only one strong leader. In order to achieve success, any great team needs multiple leaders. Consider a football team, for example. There is the head coach, and then there are the assistant coaches (offense, defense, and special teams). They may be the overall leaders of the team, but no coach has ever won the Super Bowl without a quarterback or a center. Leadership on the field is just as important as leadership off the field. You need strong offensive, defensive, and special teams leaders on the field if you intend to compete at a high level. This is called layered leadership, and it is vital to embrace and develop this concept with your team. Let's look at what happens regarding layered leadership once all units arrive on the fire scene.

We will use the example of an initial response of two engine companies and two ladder companies (three firefighters and one officer on each apparatus). The drivers of each engine will stay with the apparatus. Their job is to ensure that the interior crews quickly get adequate water. This means finding and connecting to a hydrant or other water source, such as drafting or water relays from another engine.

The first engine company arrives and is positioned just beyond the building so the hoseline can be taken off the back and carried into structure. The driver will leave room for the first-arriving ladder company, which will need to be positioned directly in front of the building. The officer of the first engine company and the remaining firefighter—the nozzle man—will advance an appropriate size and length hoseline into the house to protect the stairwell by placing the line between the fire and the occupants. They will shut the door at the top of the basement stairs and hold their position until the occupants have been removed from the building. This will help to contain the smoke, heat, and toxic gases.

If you are not a firefighter and are reading this book, right about now you are probably thinking that a fire incident is more detailed than you originally thought. You are right, and there is much more. The second engine company will help secure the first engine company's water supply (unless these companies use a bumping method and reverse jobs). They may do this by dropping their large diameter hose at the hydrant and leaving one firefighter there to help make the connection before returning to the crew. The driver of the second engine company will stay with the engine in case another water supply is needed. The officer and third firefighter will bring a second line into the building to assist in backing up the first line and to help locate, confine, and extinguish the fire. If the fire has extended to the upper floors, this hoseline may be taken directly to the second floor to help protect life. They will also ensure that the first line is fully advanced and in position prior to advancing a second, or backup, line.

Of course, those tactics may not be the right ones at every fire, but they are common practice in the industry. Tactics aside, here is a quick list of the duties that engine company personnel focus on at a fire scene. They must address and achieve the following:

- Position the engines properly.
- Secure a water supply.
- Stretch the appropriate size and length hoseline.
- Properly position the line to protect life.
- Locate the fire.
- Contain the fire.
- Extinguish the fire.
- Supply any auxiliary appliances.

- Keep their crew intact.
- Work in coordination with the ladder companies.

Both engine companies are part of the overall team, but they have specific goals that fit into the big picture, and each member is responsible to accomplish his or her designated tasks (fig. 1–6). In essence, they become a team within a team, staying true to the overall mission but operating as a separate unit. Each engine company has an officer who will act as that specific team leader (like the quarterback leads an offense). Again, this is an example of layered leadership. The overall IC could not possibly direct the interior crews the entire time. Instead, he or she directs the officers, saying things like, "I need a water supply, and two lines in the building." The IC may even say exactly where to place the lines, but when a team reaches the highest level of efficiency, this will not be necessary. The officers will already know exactly what to do.

Fig. 1–6. Every member has a specific task that must be accomplished in order to achieve success on the fireground. In the photo, Kearny firefighter Mike Kartanowicz is ensuring interior crews have a continuous water supply and sufficient pressure. *Courtesy:* Andrew Taylor.

Even to the seasoned fire service leader who is reading this book, it is impressive to contemplate the amount of teamwork needed to operate effectively on the fireground. Sometimes even the best in our industry take what we do for granted. Most civilians who stand in front of a burning house watching firefighters do their jobs have no idea what is really going on. All they know is that everyone seems to be doing a specific job.

Up to this point, we have only covered what the engine companies are doing. We have not even addressed the ladder companies. The duties of the ladder companies include the following:

- Position the apparatus properly.
- Operate and position the aerial ladder.
- Raise and position the ground ladders.
- Force entry to the structure.
- Search for trapped occupants.
- Rescue occupants.
- Ventilate the building (horizontal and/or vertical).
- Control the utilities within the house.
- Salvage savable contents.
- Perform overhaul to search for hidden fires.
- Work in coordination with the engine companies.

Each ladder company includes an officer and firefighters. The officers are also part of the layered leadership required to run an efficient fireground operation.

Clearly, a fire scene is complex. We can only excel as a team when our members know their jobs and put the mission first. In this example, everything goes smoothly. I have not addressed what happens when the incident escalates or extends to nearby exposures. In order to maintain the span of control in these situations, additional officers will need to be assigned, and the IC will constantly be assessing the situation. Given this level of complexity, the need for a high-performance teamwork is apparent. One of the most important strategies necessary to step up your teamwork can be summed up with three letters: RER.

Review, Evaluate, and Revise

RER stands for review, evaluate, and revise. These three words are crucial in the fire service. We are constantly dealing with rapidly changing conditions on the fireground. It is essential that we review, evaluate, and revise our tactics regularly to ensure we are constantly meeting our goals of life safety, incident stabilization, and property conservation. It does not matter if you are working on a hostile and intense fire scene or developing a preplan; whatever your team is trying to accomplish, the RER concept will serve you well.

The principle behind RER is simple. If what you and your team are doing is working, keep doing it. If what you and your team are doing is not working, evaluate your actions and revise them. Many teams mistakenly keep doing the wrong activity. If you are not achieving the results you want, you are either not doing the right activities or not doing enough of the right activities. Either way, you need to review, evaluate, and revise your tactics so you can begin getting the results you want.

Teams that achieve a higher-degree teamwork are great at putting their heads together and figuring out the best way to accomplish their goal (fig. 1–7). The movie *Apollo 13* includes a great scene that reminds me of what it is like being a firefighter. In the scene, a group of engineers and technicians gather to create a plan to safely return three astronauts who are stranded on a space capsule. One of the technicians dumps a bunch of supplies on the table. Picking up some of the pieces, he says, "Okay, people. Listen up. The people upstairs handed us this one and we gotta come through. We gotta find a way to make this . . . fit into the hole for this . . . using nothing but that." Without hesitation, they start looking for a solution.

As firefighters, we are constantly confronted with unusual and challenging situations. Firefighters have to concentrate on what they have to work with and solve problems quickly. If they fail to do that, the result could be serious injury or death. In order to do well, we must review, evaluate, and revise. We must adapt, which is a cornerstone of successful teamwork.

Fig. 1–7. Revising a strategy is best accomplished when people put their heads together and determine the best way to accomplish their goal. *Courtesy:* Diane Tilley.

If we fail to adapt, we fail to move forward.

—John Wooden

You Can't Achieve Significance without Sacrifice

It would be impossible to become a high-performance team without first understanding the true definition of the word *team*. A team is a group of people working together for a common purpose who must rely on each other to achieve mutually defined results. At the highest level, a team is a group of competent individuals who care about each other, are passionate about a common objective, and are fiercely committed to their mission.

There are other words that can be used in place of *team*. Some of the common synonyms that are used in the fire service are *brigade*, *crew*, *company*, *tour*, and *squad*. In some of the stories I share in this book I use a few of those terms, but for the most part, I use the word *team* to avoid confusion.

Of course, all teams are not equal. There are bad teams and there are good teams, but there are also great teams, and there is a major difference between good teams and great teams. The US Army Rangers and Navy Seals are great teams. The 1980 US Olympic hockey team and the Michael Jordan–era Bulls were great, and in my opinion, so is the US fire service. Yes, we have flaws, and I will address some of them in this book, but so do those other teams. If those teams could overcome their flaws, you and your team will be able to do the same.

There are many aspects that go into making a great team. First, the members must be well trained and prepared. The men and women who comprise your team must also be teachable and willing to work, but beyond that, they must have determination and grit. Determination requires sacrifice, and that is one area where we excel. Perhaps the greatest asset of a firefighter can be summed up in that one word—*sacrifice*. Firefighters are willing to sacrifice in a way few others are, and no team ever achieves significance without sacrifice. If you do not agree, I challenge you to come up with one example of a team that moved from mere existence to significance without sacrifice.

Sacrifice requires a generous spirit. You must be unselfish and sincerely care about the greater good of others. To win as a team, you have to believe that you are not the ultimate beneficiary. Professional sales trainers work hard to get new sales staff to understand that there will be no commission if they do not take their eyes off themselves and focus on their customers. We do not have to spend too much time stressing the importance of this with members of the fire service. The majority of men and women in the fire service who take their oaths understand their job is to put other people's needs above their own. Even so, anyone who has been in the service for more than 10 years will tell you that they have encountered firefighters, and teams, who seem to have forgotten what we are all about. It is my hope that this book finds its way into the hands of those people, and not just in the hands of dedicated individuals like you.

No team ever achieves significance without sacrifice.

Define Your Team: The Rule of Five

You are reading this book for a reason, and it is a fair assumption that you are looking to improve your team's performance to the highest level possible. One of the first steps is to define (or develop) your team. As I write these words, there are more than 100 firefighters in my department. A sales team I had developed with my part-time business had almost 20 times that number, and in both instances, I can tell you that I was able to achieve team success because of the rule of five.

The rule of five states that you are mostly influenced by the five people with whom you spend the majority of your time. This could be your five closest friends, coworkers, or family members. Those five people will play a huge part in what you do, and do not, accomplish in life and business. It reminds me of the old saying, "It's hard to soar with eagles when you are surrounded by turkeys."

There have been many studies that have shown that the behavior of your friends is contagious. An article by Mel Robbins, "The Friend Virus," which appeared in *Success* (Nov. 2010), gave the following two statistics:

- "If someone you name as a friend gets divorced, you are 147% more likely to get divorced than if you didn't have a friend who got divorced.
- If a friend becomes obese, the likelihood that you will follow suit increases by 171%."

If those statistics are true, it says a lot about how we become like the people we associate with. And if that statement is true, we should all pay more attention to the company we keep. According to Robbins, the good news is that you can use this information to your advantage if you surround yourself with people who have achieved what you hope to achieve. This could be the right attitude, wealth, a healthy marriage, professional success, or whatever else it is you are seeking to manifest in your life (fig. 1–8).

Fig. 1–8. These five dedicated individuals have achieved a high level of success within the fire service. They also happen to be friends. From left to right: Jim Duffy, Chris Peppler, Anthony Avillo, P. J. Norwood, and Frank Ricci.

If you want to succeed as a team, you have to seek out the right kind of teammates. Your close associates will play a huge role in shaping your character and your career. Parents who understand this phenomenon are concerned about who their children choose as friends. Those kids will have a profound effect on their children, and this principle holds true as we mature. We are all shaped, molded, and influenced by the people we surround ourselves with. Bad decisions, poor performance, and temptation are the by-products that result from us choosing the wrong company to keep, both in and out of the workplace.

If you get around a group of upbeat, optimistic, enthusiastic, encouraging people, they will raise you up to their level. If you spend time with negative underachievers who lack ambition and like to blame others for their failures, they will suck the life out of you. This will also transfer to the other people close to you, such as your other team members, your spouse, and your children.

We are all shaped, molded, and influenced by the people we surround ourselves with.

Life Enhancers vs. Lawn Mowers

Later in this book we will discuss ways of motivating both yourself and other various personality types. However, before we get into maximizing performance, we need to examine the influence of those closest to you. Walt Disney once wrote that there are three kinds of people: "well poisoners," "lawn mowers," and "life enhancers." I have found that well poisoners and lawn mowers share a lot of the same qualities, which can help us narrow those categories down to two: life enhancers and lawn mowers.

Life enhancers are encouragers. They are people who buy into your vision and support you, even if they do not have the drive and passion that you do. Not everyone on your team has to feel exactly like you do about your purpose, but they do have to be supporters of the very reason why your organization exists.

Lawn mowers, on the other hand, are people who do nothing but cut down your dreams and ambitions while never venturing out of their own backyards. Many people I have counseled over the years have discovered that their biggest obstacle was the people closest to them—life-sucking and energy-sucking lawn mowers.

Once you acknowledge the fact that the people closest to you are the most influential people in your life, you will begin to look at those relationships in an entirely different light. Take a moment to think about the Beatles song "Help!" and its first line, "Help, I need somebody." The second line, "Help, not just anybody" suggests that John, Paul, Ringo, and George may have understood that the people we surround ourselves with do matter.

So, what do you do if you are interested in developing a high-performance team, but you are surrounded by lawn mowers? As simple as this answer is, it can sometimes be difficult to accept. You have to change.

Your initial response may be, "*I have to change? Why me? I'm not one of the negative ones.*" That may be true, but you are the one deciding to spend time with negative people. I was in my late 20s before I understood that the people I was spending most of my time with played a huge role in determining my level of (or lack of) success. This absolutely must be dealt with before you can move forward and develop a successful team. Perhaps leadership expert John Maxwell summed it up best when he said, "The definition of a nightmare is a big dream and the wrong team."[1]

> *The definition of a nightmare is a big dream and the wrong team.*
>
> —*John Maxwell*

Surrounding yourself with like-minded people who share the same vision does not mean that you should surround yourself with clones. It has been said that if two people agree on everything, one is useless. On the other hand, if both of them disagree all the time, both are useless.

Early in my career, both in the fire service and in business, I spent time around ambitious people, but none of my friends shared the same exact ambitions that I had. I was drawn to them because they had a spark in their eye. They wanted more out of life than the status quo. However, they wanted to achieve success in different areas than I did.

I would be lying if I said the other firefighters helped me out when a promotional assessment was approaching. These assessments only occur once every three to four years, and there is never a guarantee that someone is going to get promoted. You have to hope to score well enough to come out high on the list. Then you have to hope someone at the level you just tested for retires, opening up a position that you may have the chance to fill. I do not blame my coworkers for not helping and encouraging me. I was their competition. This is also true in corporate America. The number of openings is limited, but there is an unlimited number of people wanting to fill them, which is just the nature of the game.

Luckily, several years before my promotional exam, I was introduced to an alternative way of associating with success-minded people. Through books, videos, and seminars, I was able to take my education to a higher level. When I look for new team members, individuals who are focused on personal growth go to the top of my list. You will know them. They are the ones attending the seminars and spending a lot of their downtime reading industry-related books instead of watching television all the time. Pay close attention to what people are doing in their spare time. This will tell you where their mind is at and help you define who your key players are going to be.

Great discoveries and achievements invariably involve the cooperation of many minds.

—*Alexander Graham Bell*

Communication, Coordination, and Control

Every leadership and team-building book you read will talk about the importance of communication. This one is no different. That being said, in the fire service we place greater emphasis on effective communication than most organizations could possibly imagine. When a business team fails to communicate, they risk failure. When a fire department fails to communicate, they risk far worse. Here is a true story to illustrate the point.

"I have to get out of here," the firefighter thought.

His heart was beating faster than ever before. He was on the verge of panic. It was black as midnight on the fourth floor of the rear stairwell, and wherever there wasn't smoke, there were flames coming from the floors below, trapping him on the top floor.

"I have to get out of here."

His mind was racing out of control until he envisioned his nine-year-old daughter's face. Suddenly, everything seemed to be moving in slow motion. Every morning before he left for work, he would hug and kiss that beautiful little girl. This morning was no different. He envisioned the hug they shared before he left for work, and the way she squeezed his cheeks together and said, "I love you, Daddy."

He was an experienced firefighter and did not understand how he ended up in this position, but he could not afford to think about that now. The thoughts he just had of his daughter were not going to be his last. He felt a renewed energy. He had to get out. There had to be a way.

The heat in the stairwell was becoming unbearable, even for a firefighter in full protective gear. He recognized the faint glow of daylight coming from an area where he vaguely remembered seeing a window earlier in the incident, before things changed so drastically. He crawled in that direction, reached up to confirm it was a window, and then broke it out with his axe. Peering over the windowsill, he could see a fiberglass awning 30 feet below him. He could not see any of the other firefighters from his ladder company, but he could hear them. He yelled for help. No one answered.

Through the crackling sound of fire, he faintly heard one firefighter yell, "We have to get back inside. We still have a guy up there! Tell the engine company to shut down the line!"

He contemplated jumping, but there was no guarantee he would survive the fall. It was not an impossible fall to survive, but with all the gear he had on, he was sure to be injured. Picturing his daughter's face again, he repeated, "I have to get out of here."

The other firefighters were in the rear, under the awning. They were trying to fight their way back into the building, but there was visible fire coming out of every rear window on the first three floors. They had lost communication with the missing firefighter and feared the worst.

When they initially arrived on scene, before the situation deteriorated, the ladder company had entered the four-story apartment complex from the rear because they were told people were trapped inside. The first-floor fire was too heavy in the front of the house, and the first engine company had not yet arrived. Their plan was to enter through a rear stairwell and search where they could until the engine crew was able to get water on the fire. The plan seemed to make sense at the time, but once they reached the second floor, the fire traveled from the front of the building to the rear and began moving up into the stairwell.

The speed at which the fire was traveling did not make sense to them. They were overcome quickly, leaving no time to think, only to react. They had to get out, otherwise they would burn up. They thought the entire team was able to get out, but the fire cut off one firefighter's escape, and he was forced to go up the stairwell instead of down and out like the others. They felt responsible, but the fire was heavy and furious. There was no other option.

Outside, they frantically tried to communicate with the engine company to tell them to shut the line so they could attack from the rear and protect their trapped team member with a hose stream. Their requests were not being heard. The engine company was engaged in an intense battle and could not make out the radio transmissions.

Alone, and as close to hell as he could imagine, the trapped firefighter was hoping that the others below would see him through the column of smoke and raise a ladder to the window so he could get out. He never considered the fact that it was a hopeless plan. Between the awning and power lines in the rear yard, it was impossible to raise a ladder. Things quickly went from bad to worse. Without warning, the fire began ferociously climbing up and surging toward him. There was no other option. He exited the window head first, but first he hooked his right arm firmly around the windowsill. He swung his legs out and hung there.

"He's coming out the window!" someone in the rear yard yelled, alerting the others.

"They see me," he thought. It was great news for a second. Then he remembered they still could not get to him and yelled out, "Help me!"

They wanted to help, but there was nothing they could do. The firefighter was now hanging by his gloved hands. The flames were blasting out of the window above his head. His arms stretched out. The sleeves of his turnout coat stretched, exposing the flesh around his wrists. He could feel his skin burning. The pain was intense. He would not be able to hold on much longer. He felt his grip slipping. Seven feet below him, and to one side, was an air conditioner protruding from the wall. He swung and stretched his leg out and tried to touch it with his foot. It was too far. He could not hold on any longer.

The firefighter fell 30 feet and crashed through the awning, shattering the supporting two-by-fours. He landed hard onto a cement patio, barely missing a discarded washing machine and a table that was upside down. Six inches to the right, and he would have been impaled on one of the table legs. Six inches to the left, and his head would have slammed into the washer. He was badly injured, but he was alive. He would get to see his daughter, and that was all that mattered to him.

This story happened to a brother firefighter of mine from a neighboring town. To someone outside the fire service, it might appear as if the fire was the reason why he became trapped. There is much more to the story. The firefighter was forced up the rear stairwell because an engine company had entered the structure and began putting water on the flames, opposite his company working on the other side of the fire. The wind behind the engine company, along with the direction of their attack, pushed the fire toward the rear of the building and created the dangerous situation. It happened, in part, because the engine and ladder companies did not properly communicate with each other.

It is not sufficient for a three- or four-member company to work as a team. When working at a structure fire and any other large-scale incidents, this company is only one component of the team. The actual team is comprised of every individual and company that responds to a call. There are three key components—the three Cs—that will help your team operate as a cohesive unit and avoid this type of outcome on your fireground: communication, coordination, and control. Here is how the three Cs are applied in the fire service (fig. 1–9).

Fig. 1–9. Communication, coordination, and control are key components that will help ensure the success of a team. *Courtesy:* Ron Jeffers.

Communication

At least one member of every company has a radio, but having a radio and using it are two very different things. Each team leader communicates and provides progress reports regularly so the IC and all members on scene know what every other company is doing. In order for this to work, messages that are transmitted on the fireground are often kept to a minimum, and only essential bits of information are shared.

Coordination

Engine company personnel must know where ladder and rescue company personnel are located within the structure. The only way to avoid situations like the preceding story is through coordination. When an engine company opens a door to advance a hoseline or perform a search, or a ladder company opens a window to ventilate, they are going to change the environment and possibly redirect the flow path, which can push the smoke and heated gases to other areas within the structure. They cannot afford to be so focused on their job that they forget to take a second to think about how their actions will affect the overall mission of the team.

Control

When you are part of a team, you must remain in control of your actions. When stress and anxiety levels are high, it is easy to lose control and panic. As firefighters, we train constantly to ensure this does not happen. It is important to stick to the game plan. If a problem or urgent situation arises, communicate the message with the IC or operations officer so that person can coordinate with the other teams working on scene and get you the help you need (fig. 1–10).

No team will achieve and maintain true success without communication, coordination, and control.

This is not a new or original concept. We all know we need to communicate, coordinate, and be in control. Athletic teams know it, sales teams know it, and educational institutions know it. Why, then, is it so hard to establish and maintain these three essential components?

The answer is lack of discipline.

Fig. 1–10. Tactically persuasive strategy

Knowing how to do something right is easy. If you are a healthy person with two arms, you know how to do push-ups. You also know they will help you become stronger. You can also imagine how doing only 100 push-ups a day can improve your overall health and strength. So, if you know how to do it and why you should do it, why aren't you doing it? The answer is that the hard part is actually choosing to do it and repeating that activity until you achieve the results you want. Knowing how to do something right is the easy part. Having the discipline to do it is the hard part.

What Stage Is Your Team In?

In the fire service, just as with any other industry, there are people who are good at what they do, people who are great at what they do, and people who may not be cut out for the job at all. No two individuals are alike, making it tough to generalize. However, for simplicity's sake, consider placing each of your team members into one of three categories we refer to often in our industry: probies, veterans, or dinosaurs.

A *probie* is a new (probationary) firefighter who may be green but is willing to learn. Probies have been through the academy and are beginning to understand what they need to do in order to become productive team members. A firefighter's probationary period is usually the first 12 months, so he or she may not have actually fought a real fire other than controlled academy fires. With the right guidance, probies can become highly productive and valuable assets to your team.

Veteran firefighters are seasoned. These individuals have been through their share of real-life incidents. They have held the nozzle in the fire, treated victims at vehicle accidents, and experienced a variety of situations that helped them develop their skills, confidence, and competency. They can be your greatest assets, but with the wrong attitude, they can also become your greatest obstacles.

Dinosaurs can be the biggest challenge of all. The probie is eager to learn, the veteran is still contributing, but the dinosaur is done learning altogether. They simply do not care anymore, or they do not care enough to change. These people can cause a serious problem to your entire organization and have a negative effect on your team's overall attitude. (Note: don't confuse the term dinosaurs with age. A person who still cares and contributes is a veteran, not a dinosaur.)

One of the goals I have always strived to achieve is to develop an atmosphere of constant growth and improvement. It is amazing how dinosaurs tend to avoid joining teams that are proactive. All three categories have their challenges, but dinosaurs seem to provide the most resistance for leaders who are trying to improve their team's overall performance.

Before you can improve, you should first determine your team's current stage of development. By most standards, and according to most publications, there are four stages of a fire: incipient, growth, fully developed, and decay (fig. 1–11). When you observe how a fire progresses and eventually burns out, you can easily see how you can also use these same four stages to help determine at what stage your team is functioning with regard to developmental progress.

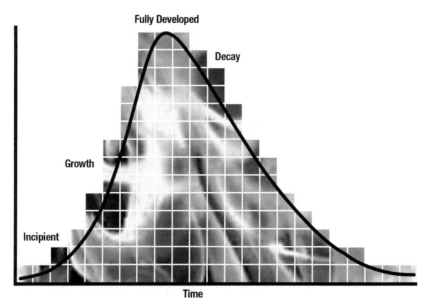

Fig. 1–11. Stages of team development

Traditionally, a team will go through these four stages of development. Each stage presents its own special challenges to a group of people striving to work together successfully and forming a cohesive team. It is important to identify where your team stands, so you can determine which actions need to be taken to support the team's quest for success.

Stages of team development

I will begin by explaining what each stage is in terms of fire development. This will help you understand how each can be categorized as a team development stage.

Incipient. This first stage of fire begins when heat, oxygen, and a fuel source react chemically, resulting in fire. This is also known as *ignition* and is usually represented by a very small fire, which sometimes goes out on its own before the subsequent stages are reached. Recognizing a fire in this stage provides your best chance at suppression or escape.

In terms of team development, the incipient stage occurs when a group of people come together with intentions of accomplishing a shared purpose. The team may be small, but they have ignited and are ready to produce. In this stage, you usually have a combination of probies and veterans. Dinosaurs may be on the team as well, but since they do not like change, they are usually not contributors in a positive sense. When these teams burn out on their own, like some incipient fires do, it is usually a result of the negative energy of dinosaurs or the inability of people in leadership positions to encourage and inspire team members.

Growth. The growth stage is where the structure's fire load and oxygen are used as fuel for the fire. There are numerous factors affecting the growth stage, including where the fire started, what combustibles are involved or are nearby, room size, ceiling height, and the potential for thermal layering. It is during this shortest of the four stages when a deadly flashover can occur, potentially trapping, injuring, or killing firefighters.

Relating to team development, in the growth stage, results start to happen, and a team will begin to develop and produce. They may also encounter challenges, such as disagreements about mission, vision, and strategy. During this stage, team members are getting to know each other. Strained relationships and conflict may occur. This may be the result of clashing personalities, but it can also be the handiwork of the dinosaur mindset that change is not good. This is a critical stage. With the right energy, philosophies, and guidance, it can lead to the most important phase.

Fully developed. When the growth stage has reached its maximum and all combustible materials have been ignited, a fire is considered fully developed. This is the hottest phase of a fire.

Relating to team development, the fully developed stage occurs when the team has consciously or unconsciously formed working relationships that are enabling progress on the team's objectives. Relationships, team processes, and the team's effectiveness in terms of working on objectives are syncing to bring about a successfully functioning team. This is where momentum occurs and you have the best chance of reaching your team's autoignition point. You will know you have reached peak performance when the team is functioning so well that members believe it is the most successful team they have ever been part of.

Decay. Usually the longest stage of a fire, the decay stage is characterized by a significant decrease in oxygen or fuel, putting an end to the fire. There are two common dangers during this stage. The first danger is the existence of nonflaming combustibles, which can potentially start a new fire if not fully extinguished. The second danger is the possibility of a *backdraft*, which occurs when oxygen is reintroduced to a volatile, confined space.

Relating to team development, in the decay stage, morale is low. Energy levels have significantly decreased, and the team seems distracted, unfocused, and consumed with drama. This is the most dangerous stage for a team; everything you have worked to build is on the verge of collapse. You can come back from the decay stage, but it will take focus, hard work, and a united effort. It will be necessary to focus on passion and purpose to get the team back in the game.

Teams do not necessarily move through these stages in order. Often new team members (probies) can be the catalysts that will help the team move quickly from one stage to another. This can be a good or bad thing, depending on the energy, experience, and desire these members bring to the team. A dedicated group of veterans and a focused leader will help direct the energy of all members in the right direction. The goal of all team members should be to avoid the decay stage at all costs, because it is the most difficult stage to survive.

Teamwork Strategies

Strategy and *tactics* are two words every firefighter must know and understand. Without sound strategy and tactics, we could never achieve our goals at a structure fire, or any incident for that matter. This book provides you with specific tactics you can implement to help your team succeed. (See the "Team-Building Exercises" section as an example.) At this point, however, we will take a moment to talk about strategy.

A *strategy* is a high-level plan, usually implemented by team leaders, which is devised to achieve one or more goals under conditions of uncertainty. Strategy requires planning and thinking because the resources available to achieve these goals are usually limited. Strategy is not about choosing one plan and sticking to it no matter what. It is more about attaining and maintaining a position of advantage and putting your team in a position in which they can adapt if needed and still accomplish your overall goals.

It is difficult to generalize strategy for teams without knowing where you are and what you are currently trying to accomplish. However, there are some absolutes to keep in mind. Implementing the following seven strategic steps will elevate the level at which your team operates.

Seven strategic steps to elevate team performance

1. Emphasize group recognition. It is blatantly obvious to others when a team leader "plays favorites" or treats some people unfairly. Focus on distributing credit to the team, not just one or two members who may have outperformed the others. It is okay to mention and point out great behavior and accomplishments of individuals; it shows others what you are looking for. However, do not overlook the contributions of the rest of the team.

2. Reward excellence. There is a difference between a base hit and a grand slam. When your team hits it out of the park, make a big deal out of it. Do it in a way that makes the members want to continue to work at the high level of efficiency needed to achieve those results. When excellence goes unrewarded, people become frustrated and start to look for appreciation elsewhere.

3. Encourage and promote clarity. The team needs to know what its job is and what the end goal is. In order for this to happen, leadership must be transparent, and the members need to be involved in the goal-setting and decision-making processes. When everyone on the team pitches in from the early stages, they get connected in a way that makes them feel like they own the mission. When everyone on your team is on the same page and they take ownership, you are on the verge of achieving true team success.

4. Mitigate conflict. Rumor, gossip, and cliques that hold secret meetings can destroy a team. Do not allow conflict to grow to the point where it divides team members; stop it early. When your team members are consumed with drama, it is a sign that they are no longer connected to the goal. Their time and energy is now spent on the wrong activities. Conflict resolution is a skill that must be learned, but stopping conflict before it gets out of control is an art that will benefit your team in ways you could never imagine.

5. Equip them properly. When people are assigned a task or job, they will need the education, time, and tools to get the job done (fig. 1–12). This is sometimes tricky, because the concept of strategy comes from having limited resources. Limited resources require training, planning ahead, and thinking. At the very least, help establish an environment where team members can use whatever resources they have available to them, without distractions.

Fig. 1–12. Do not assign a task unless your team members are equipped with the right training and tools they need to get the job done.

6. Fix what is broken. When you make a mistake the first time, it is a learning experience. When you make the same mistake a second time, it is a choice. The goal should always be to find the *one best way* to accomplish any specific task. In order to do this, we need to break old habits and try approaching tasks from different angles. If a certain tactic did not produce the results you want, change it.

7. Cross-train. A common mistake made in the fire service occurs when people get locked into certain duties. It is important to master one or two tasks, but it is also common in smaller departments (prevalent in most of the United States) for a member on a ladder company to be told to do an engine company duty, or vice versa. That being said, each member of your team should be trained to do a variety of tasks. This way, your team members will become interchangeable. If one person is not available on a specific day to accomplish a specific task, you will be able to use another person in that position. You still want to allow individuals to run in their lanes and work with their strengths, talents, skills, and abilities. However, promoting interchangeability will benefit your team in the long run.

Without the right strategic plan, you cannot create momentum. Without momentum, you will never become a high-performance team.

Creating Momentum

I received an e-mail one day from a newly appointed lieutenant halfway across the country, asking if I could give him advice. In short, he said he was younger than most of the others in his department and was concerned they lacked passion and desire. He was an ambitious young man who did not want to make waves but wanted to make a difference. He wanted change, explaining that he wanted to create the kind of momentum that his organization had when he first joined the department. We e-mailed back and forth for several weeks as he worked to achieve his goal, which he did with great success, to his credit.

Firefighting is a service-driven (not sales- or production-driven) industry. Nevertheless, creating team momentum is vital to the fire service. No matter what type of team you are on, one of the keys that will determine your level of success is momentum. Merriam-Webster defines momentum as "strength or force gained by motion or by a series of events." Momentum will help your team reach great heights. On the other hand, the lack of momentum, or inability to create it, can leave your team feeling unmotivated, stagnant, and hopeless.

What does momentum look and feel like? In his book *The 21 Irrefutable Laws of Leadership*, John Maxwell says, "Momentum is really a leader's best friend. Sometimes, it's the only difference between winning and losing."

You will know you have momentum when your team no longer sees problems as anything more than opportunities. With momentum, you effortlessly go over, under, around, or through obstacles that would normally have stopped your team dead in its tracks. You build from success to success and are able to adapt and overcome with ease. When this phenomenon occurs, people throughout the organization are motivated to achieve more, and at a higher level. If you have not experienced it before, trust me, you will know you have team momentum the moment you reach it.

You will know you have momentum when your team no longer sees problems as anything more than opportunities.

Team leaders love momentum because it makes them look better than they actually are. When we respond to multiple fires a week, our team can get in a groove and begin to perform at a highly efficient level, called *performance momentum*. The opposite can occur when we go months without a significant working fire. I am proud of the administrative staff of my department in achieving the performance momentum level in grant writing. During a five-year period, it seemed like we were awarded every grant we applied for, reaching a combined total of $3 million. We achieved every goal we set, and I am proud to have been a member of the team during that time.

With fires, it is difficult to create momentum. They happen when they happen. With that in mind, the worst thing your team can do is to wait for things to happen. The better option is to train hard and often and work toward creating momentum in terms of team efficiency. Instead of focusing on what you cannot control, let's focus on what you can control.

Here are eight steps to help you create momentum in an existing team.

Eight steps to creating momentum

1. Desire change. Momentum usually starts with one person and then moves outward and impacts the entire team. To make real change, a person must have passion. Leaders are ultimately responsible for generating momentum on the team. They do this by implementing action steps to show others that they are creating personal momentum. If you cannot motivate yourself, you will not be able to motivate others. Once you take charge of your personal actions and set the example for the team, expect others to follow. The speed of the leader often determines the speed of the team.

2. Introduce something new. I find that officers are at their best when a new firefighter is assigned to their group. Firefighters are at their best when a new challenge is presented, and they begin to focus on a solution. Often something new triggers momentum. A new tool, a new training prop, a new goal, a new team member, a new officer—all are examples that can provide an extra degree for your team (fig. 1–13). Many companies come out with a new product to create momentum. Look at what happened to Apple when they introduced the iPhone. The concept of introducing something new also applies to a new way of thinking. Keep in mind that you will never solve the problems you are experiencing with the same thinking that put you there in the first place.

3. Set goals and work on improving. Momentum is sustained through consistent improvement. What is your team improving on right now? Develop a clear vision and goals for your immediate future. Clarity is essential. Even if you do not have the exact action plan, once you know the general direction you want to go, you can begin leading your team in that direction. Continually remind yourself and your team about the vision you are pursuing. Put them on paper and post them where people can see them daily (fig. 1–14). When the team is focused on the goal (the prize), they will be able to overcome the inevitable struggles and conflicts that stop most teams dead in their tracks.

Fig. 1–13. My organization acquired abandoned buildings that were scheduled to be demolished and used them to train on skills like roof ventilation. These drills helped us create momentum.

Fig. 1–14. Develop a clear vision and goals for your immediate future and put them on paper.

4. Generate forward progress. Achieving momentum starts with creating forward progress. Getting started is the most difficult part. You already know what inertia is—objects in motion tend to stay in motion. And you also know that objects at rest tend to stay at rest. As a leader, make it your mission to be that outside force that gets a team to begin taking steps toward your goal. The way you start your day is important for this very reason. If you begin your day by immediately picking up where you left off the day before, your team will get in the habit of doing everything with *specific intent*, with every action designed to achieve a specific result. This way of thinking will help your team generate forward progress.

5. Create a sense of urgency. The speed at which you pursue your goals affects everything. Setting deadlines helps create a sense of urgency. Like your goals, you will want to post your designated deadlines where everyone can see them, such as on a large whiteboard in the room where your team spends most of its day. Apply the principle of massive action, in which a massive amount of specific action steps are compressed within a short amount of time. Teams will not always have to work in an urgent manner, but when you are trying to achieve momentum, creating a sense of urgency is a great way to do it.

6. Run a competitive promotion. Teams like to compete against other teams, especially when there is a prize at stake. Set a goal that everyone can run for and offer rewards for those who produce the best results. Make it fun and provide recognition for all who compete. I led one particular sales team for more than seven years. Our company offered promotions on a quarterly basis. My team never won a promotion, but we were always in the top five nationally. We were also the top-producing team in New Jersey, and one of the top in the northeastern United States all seven years, because we participated and competed to win every promotion. Any organization can run a promotion that offers prizes for teams and individuals who do the right activity. The right activity, over time, creates momentum.

7. Build from one success to another. I have found that the best time to run a team drill is right after another one. The best time to fight a fire is right after another one. The best time to set a goal is right after you just achieved one. Why is this? We are at our best when we build a rhythm. One of the worst things a team can do is taste success and follow it up by taking a long break. Use small successes as a motivator to help propel your team to bigger things. Your team should be highly motivated immediately following a success. Take advantage of that motivation to continue the forward motion you have started.

8. Celebrate victories along the way. Although you want to move from success to success, you should still find ways to celebrate along your journey. A football team may have scored a touchdown, but that does not mean they have won the game. They should, however, celebrate the touchdown, because if they score enough times, they will win the game. Rewarding your team for a job well done can be as simple as taking a break and surprising the team with cake (I know my health-conscious friends just cringed). If members of your team perform at an exceptionally high level, give them a round of applause. We do it for people who entertain us, so why not do it for those who make us a better team?

A small body of determined spirits fired by an unquenchable faith in their mission can alter the course of history.

—Gandhi

Team-Building Exercises

Firefighting *is* teamwork. Athletic teams do not just show up on game day and expect to play well without preparing. They do things that build their ability to work, react, and perform together. We need to do the same. It is not enough to simply eat together as a crew, although that helps. We have to take it further than that. It is not enough that team members just show up; they must show up with the specific intent of improving. We have to utilize our time together and deliberately choose activities that enhance our ability to excel as a team.

An unfocused team will have more downtime than a focused one. This is partly because they lack goals and partly because they lack creativity. There are many ways to enhance your team's ability to work together and perform at a higher level. Below are 10 team-building exercises I have found to be productive. Try them out. I share these exercises from a fire industry perspective, but if you are looking to develop a team outside the fire service, you can use the same concepts and apply them to your chosen field.

Real-time drills

If you have ever attended one of my seminars, you have heard me talk about the benefits of real-time drills. In fact, I believe in this concept so much as a team-building exercise that I listed it first. The idea for real-time drills as a team-building exercise came by accident. We were performing an inspection of a supermarket one day when I asked the six firefighters who were with me how long they thought it would take to raise the aerial ladder, access the roof, and cut a ventilation hole. I received various answers that ranged from 2 minutes to 15 minutes. Two minutes seemed unrealistic to me, and 15 minutes was unacceptable. The thing that struck me the most, however, was that they had no idea how long it would take.

This presented a very big problem. If firefighters inside a structure were struggling to put out a fire, and needed quick ventilation to remove the heated gases and smoke from the structure, not knowing how long it would take to accomplish this task could cause us to misjudge the activity. The ultimate result could be a failed mission and serious injury or death. I pondered the answers that I had received and wondered how I could best approach this problem. Then I remembered that my brother had once organized a drill where he timed firefighters as they accomplished a similar task.

About two hours after the question, I returned to the supermarket, called my dispatcher, and requested that the ladder company respond back to the scene. They arrived a few minutes later, unsure of why they had been called.

When the officer exited the apparatus, I said, "Here's the scenario, Cap. We have a working fire that is getting ahead of us. I have an engine company that is being pushed back out of the building. We need the roof opened up now. Have your guys do exactly what they would do. When you get on the roof, start the saw and simulate the cut. I'm timing you, and the clock starts now."

I could tell the captain had questions, but I walked away, keeping my eyes on the stopwatch, so he turned to his crew and gave them instructions, and they went to work. It took more than 5 minutes to access the roof and start the saw. Within 7 minutes, they had begun simulating the cut. I told them I was adding 1 minute as a buffer, because it would take time to lift the roof and clear the hole. The end result was 8 minutes. Now we knew exactly how long it would take to accomplish this goal at our current skill level. This was the introduction of the real-time drill.

After we completed the exercise, I had our other ladder company respond to the scene and do the same. The only difference was I told them that the previous company had accomplished the task in 8 minutes. Their competitive nature took over, and they beat the first crew's time by 30 seconds. After that, we set the goal to have both companies reduce their time through practice over the next 30 days. The real-time drills became vital to our success as a fire department. We took that experience and expanded into other areas of responsibility, conducting real-time drills on vehicle stabilization and extrication (fig. 1–15), engine company relay pumping, self-rescue techniques, and more.

The theme of the drills could be summarized in three words: now we know. Now we know how long it would take to breach a wall. Now we know the speed at which we could stabilize an overturned vehicle and remove the doors. Now we know how fast each individual could connect 50 feet of 5-inch hose to a hydrant and secure a water supply.

The biggest benefit that comes from real-life drills is getting a company to work together toward a common goal. They become a team, competing against another team for bragging rights, while realizing that the training benefits the organization as a whole.

Fig. 1–15. Kearny firefighters participating in a real-time stabilization and extrication drill.

Practice scenarios

Have you ever stopped to think about how much time firefighters spend around the kitchen table at work? I am not talking about time where we are eating breakfast, lunch, or dinner, but the time before and after meals, and the time between drills and housework.

One of the signs of being a well-organized and proficient team leader is that you know how to utilize downtime in a way that can benefit the team. One way to do this is by discussing scenarios. The concept is simple: present a scenario and discuss what your team would do to solve the problem or mitigate the situation.

This can be accomplished by simply printing out photos of various types of incidents. For example, place a photo of a structure with smoke or fire coming from it on the table. Go around the room, asking questions such as the following:

- What size-up factors would influence our strategic decision making?
- Where would the first line go?

- What assignment would you give the first-arriving ladder company?
- Are there any signs of people being home?
- What is the best way to ventilate this building?

Do not stop at the structure fire. You can use the same technique with building construction, hazardous materials incidents, signs of collapse, rescue operations, and just about any emergency you can think of. This exercise will not only get your crew thinking like a team, it will also give you a good idea about how the people around you think. You will discover that some individuals provide well-thought-out answers. One of the most valuable assets on the fireground is a firefighter who thinks like an officer. This is the type of exercise that will help you see the early signs of a leader developing.

Another way to do this is exercise is through the written scenario. Instead of using a photo, give a written scenario to your frontline leaders that revolves around supervisory challenges or administrative tasks. For example, provide a subordinate challenge to the officers and discuss what actions you would take to resolve the problem. This will get your members thinking at a higher, more responsible level.

Physical challenges

Team building does not always occur in the firehouse or on the training ground. It can also happen on a Saturday afternoon, knee deep in mud. Firefighting is a physically demanding job, yet many organizations throughout the country do not have formal fitness standards that individuals must maintain. Because of this, your team has a choice. They can choose to ignore the facts, or they can set their own fitness goals.

What are the facts? According to a Harvard School of Public Health study, firefighters are 300 times more likely to suffer a heart attack at work than any other profession.[2] Heart attacks are the number one killer of firefighters. In fact, heart disease kills more on-duty firefighters than anything else, and it is definitely linked to their emergency duties. The study states that the most frequent cause of death among firefighters is coronary heart disease rather than burns or smoke inhalation. Cardiovascular events account for 45% of deaths among firefighters on duty in contrast to 15% of all deaths that occur on conventional jobs.

Some possible explanations for the high mortality from cardiovascular events among firefighters include smoke and chemical exposure, irregular physical exertion, and the handling of heavy equipment and materials. Other factors are heat stress, shift work, a high prevalence of cardiovascular risk factors, and psychological stressors. According to Harvard researcher Stefanos N. Kales, MD, "Battling fires and rescuing civilians involve sudden, extreme exertion and exposure to toxic environments. These job risks make heart disease all the more deadly for firefighters."

So, if firefighters are 300 times more likely to suffer a heart attack, and they have a 100% increased risk of cancer, 50% higher risk for multiple myeloma and non-Hodgkin's lymphoma, and 28% higher risk of prostate cancer, why aren't we challenging our team members to do something about it?

Grab a team of coworkers and have them join you at fun events like mud runs, military-style obstacle courses, and even zombie runs (yes, they actually have zombie runs). Events happen throughout the country all year around (fig. 1–16). You can choose an event, a team name, a logo and motto, and you can even design your own shirts. You can train together, support each other leading up to the event, and work together to get through the course on the day of the challenge. At challenges like these, you will learn the value of strength in numbers, and the importance of working smarter instead of harder. You will also see how far you can push yourself and the team, and you will more than likely be surprised in all areas.

If these types of physically grueling challenges are not right for your team, consider going fishing or whitewater rafting, or organize a game of softball or paintball. Choose some type of activity where you can form a team and compete. We are in a physical profession, so why not choose a physical team-building exercise? Fitness challenges (or any of the other activities listed above) are a great way to have fun, build morale, and help a team develop great habits.

Step Up Your Teamwork

Fig. 1–16. Physical and military-type challenges like the GORUCK provide great opportunities for individuals to pull together and work as a team.

Community relations demonstrations

Consider how many events occur in your community on a yearly basis. If your community is like mine, you have a variety of public relations opportunities that you may not be taking advantage of. These could include graduation ceremonies, proms, block parties, parades, community picnics, carnivals, volunteer appreciation events, wet downs, farmers markets, New Year's festivals, fireworks displays, holiday parties, and Memorial Day and Veterans Day celebrations, among others.

Any time a group of people in your community gather is an opportunity for you to improve community relations. How can you do this? You can start by handing out free giveaways, like fire prevention handouts, coloring books, and home safety tips brochures. If your apparatus or some equipment is new, you can show it off for the hard-working taxpayers to see. These things are great, but there is something special about performing live demonstrations for the public to witness.

One of the favorites in my organization is an extrication demo. Throughout the years, we have performed live demos at several events.

Perhaps the most gratifying ones are when we perform them in front of the local high school before prom weekend. There is an epidemic of texting and driving sweeping America. It is as deadly, if not more deadly, than driving under the influence. We can talk to teenagers until we are blue in the face about the dangers of all three (texting, drinking, and drugs) as it relates to driving, or we can simply show them (fig. 1–17).

Fig. 1–17. Planning and conducting community relations demonstrations like the one shown in this photo will not only help your team work together but also may save lives.

The reason why this exercise is so gratifying to many of our members is because they can tell they made their point by the look on many of the students' faces. Planning a community relations drill like this takes a lot of effort. When we do it, we contact a local towing company to get vehicles to practice on a week in advance. Then we coordinate with law enforcement officials on ways to make the greatest impact. Our usual format is to set up a simulated vehicle accident, with multiple wrecked cars, then have the school officials call all the students outside to the front of the building. The scene is set, complete with student volunteers who pose as victims.

Once the students are ready, we pull around the corner with lights and siren on, exit the apparatus, and begin performing the extrication. A sign is usually set up saying something like, "Stay Alive, Don't Text and Drive." We also designate an individual to say a few words over a PA system to make sure the point is clear as we package and relocate the victims from the ground to the back of the ambulance. We always make it a point to cover one individual completely, just to make the greatest impact. We are in the business of saving lives, and so much of this can be done in the prevention stage. What could be a more noble team-building exercise than developing a community relations demonstration that may actually save lives?

Multi-instructor drills

I first organized a multi-instructor drill when we were conducting technical rescue refresher training. I was set to do the training myself when it occurred to me that I had four individuals on my shift who were better than I was at each of the four stations we were training on. One was great at making Z-rigs, another was great with ropes and knots, a third had an excellent understanding of our confined space communications system, and a fourth was well versed with setting up the tripod system.

I am a big believer in the many advantages of getting other people involved and utilizing the talents of the people around me. This presented a great opportunity to let others shine. After discussing the idea with each of these individuals, everyone became excited. We set up four stations and had groups of four firefighters move from one station to the next, until they all completed each discipline. The teaching process was simple. The station instructor would explain the lesson, demonstrate the technique, do it along with the team, and then let the team do the exercise alone. The results were great. No one became bored, which often happens when you listen to just one person for several hours. Everyone learned something new, because they were listening to new instructors. Additionally, the four individuals who were instructing that day (two firefighters and two captains) gained a greater sense of purpose and pride. They contributed and felt great about their part in the drill.

When you plan multi-instructor or multistation drills, you enlist help from several team members. In the process, you will see some of them rise and perform at a higher level because they are not just participating in the drill—they are part owners of it.

Full-scale, multiagency drills

Think about all the working parts of a full-scale, multiagency incident—the unified command, interoperable communications, command staff designations, incident operations, strategy and tactics, company assignments, coordination, goals, and objectives. Just the prefix *multi* suggests that you have many moving parts.

This is why designing and organizing a full-scale, multiagency drill can be a great team-building and learning exercise for all involved. When you participate in a drill like this, you get to practice your portion of the mitigation process. When you design a drill like this, you see the overall picture.

I remember going with some other firefighters to see the New York Rangers take on the New Jersey Devils at Madison Square Garden. Our seats were two rows behind the Devils players. I was amazed at how fast the players were, but I missed a lot of the action because we could only see the players who skated right in front of us. Before the start of the last period, one of the guys pointed to the upper tier and asked if we had ever watched a game from above, saying that it was also fun to watch from up there. My response was, "Why would you want to watch a game from all the way up there after being this close?" He explained that you could see the plays form and enjoy the game from an entirely different perspective. After hearing him talk so passionately about how great it was, we left our seats and found three empty seats in the upper tier to finish watching the final period of the game. Although I preferred the closer seats, I had to admit, it was pretty cool watching the formations from a higher vantage point. It really was an entirely different perspective, and if I were a coach, I would want to watch tapes of my team from that view so I could better prepare them to work as a team.

This is the same feeling you will have when you and your team design and plan a multiagency drill. You will get an entirely different perspective. You will understand the importance of coordinating as you develop the drill. You will develop relationships with representatives from various organizations, and you will have the opportunity to expand the influence and credibility of your team beyond your internal organization. Take the initiative to sit down with your team members, come up with an idea for a drill, and contact other organization leaders. It is a great team-building exercise.

When a team outgrows individual performance and learns team confidence, excellence becomes a reality.

—Vince Lombardi

Build a prop

Fire departments are made up of talented individuals, many of whom are skilled carpenters who can swing a hammer with the best of them. Why not take advantage of the talents, skills, and abilities of your team members and use them for the betterment of the organization?

Think about the many props your organization can use. Things like rescue dummies, obstacle courses (for mask confidence and self-rescue drills), and confined-space training boxes. You can build props for forcible entry (where you can practice forcing doors), and vertical ventilation (peaked roofs that are built close to the ground so you can practice cutting holes).

Years ago, a handful of Kearny firefighters built two community relations props that we still use many times throughout each year. They are wooden props designed to look like window frames. In each window is fire which is also made of wood and placed on hinges (fig. 1–18). When we attend events like grand openings of businesses, carnivals, and 4th of July celebrations, we bring the props and watch the children excitedly line up and await their turn to direct the stream at the fire in the windows (fig. 1–19). We also get more than a few adults in the line. Even our mayor and council members took a turn on the prop one year.

This was a simple prop to build due to the team's dedication, careful planning, and ability to work together.

Figs. 1–18 and 1–19. Our members enjoyed building these props, which we use several times a year to create memorable experiences for our younger public at community relations and educational events.

Our members also designed an incredibly elaborate technical rescue drill scenario using multiple props. The drill consisted of a 40-foot vertical rescue, a 30-foot horizontal rescue, and a 20-foot angled rescue. All training stations were designed and built by our members using large-diameter sewer piping and standard construction materials. Team members secured grant money to help offset the costs, located a warehouse and obtained permission to use it to train in for a month, and lined up experts to provide our members with advanced technical rescue training.

Many organizations enjoy the challenge of developing a firefighter skill obstacle course. These props enable firefighters to practice skills like reduced profile or SCBA quick-release techniques, freeing yourself from entanglements, and using a hoseline to help find your way out. The thought of becoming trapped, disoriented, or lost inside a smoke-filled environment can be terrifying, so this exercise will do much more than get the team to work together—it can save firefighters' lives. It is not difficult to find ideas on how to make a firefighter survival or mask confidence course. A simple Internet search on credible sites will bring up at least 100 pages that provide pictures, tips, and ideas for developing one.

Do not underestimate the advantages of bringing a group together to design props like those listed above. Doing so requires a commitment to develop the organization, which always helps improve morale and team spirit.

Culture development exercise

One of the best team-building exercises you could possibly do is one that will help your team create an organizational culture. My favorite way of doing this is by taking a large drawing tablet or dry erase board and making three columns. At the top of the columns, write the words Keep, Stop, and Start. Once this is complete, begin discussing the following questions:

1. What are we doing well that we need to keep doing?

2. What are we doing incorrectly that we need to stop doing?

3. What are we not doing at all that we need to start doing?

You will be amazed by how in touch your members are with right and wrong activities. This exercise simply provides you with a vehicle to have deliberate conversations about what is important to your organization. Discuss all aspects of your job, from performance in the field to providing exceptional customer service. This is also a great time to come up with a team name and mission statement. When you and your members fully understand what activities you need to keep doing, stop doing, and start doing, you are on the way to team success.

Write a grant proposal

Every organization has needs. Maybe you have outdated turnout gear. Maybe you want to start a technical rescue team and need the training and equipment. Perhaps you need a new engine or fire boat. Maybe you need a fit testing machine, thermal imaging cameras, a laptop and multimedia projector, or new multigas meters. You may even need 16 new firefighters to bring yourself up to NFPA standards. I mention these specifically because those items are among the successful grant proposals that my department has been awarded.

The grant writing process can be one that involves many people for a good cause. Tens of millions of dollars are given away to fire departments through grant programs in the United States alone. The money is there for the asking, but writing a proposal is a very competitive process that favors organizations that use the talents of more than one individual.

Begin by forming a committee and identifying a list of needs and a type of project you want to fund. After you do that, look for grant programs that you are eligible for. Once you find them, read the guidance and determine if what you are looking for is on the highest priority list of that grant program's goals for the year. It is important to know and understand the process, and the best way to do this is by getting others involved.

After you have identified your project, you can start the application process. Some members may attend a grant support workshop or do research online. Others may identify costs of equipment and training. Others may come up with a list of standards your organization is not in compliance with but would be if awarded the grant. Others can collect information such as call volume, types of incidents you responded to over the past several years, or community population and demographics. All of this, coupled with a strong narrative explaining why the project should be funded, will help you be successful in acquiring grant money.

As you can see, a lot goes into researching, preparing, and submitting a grant proposal. By getting a team together, you increase your chances of making a compelling case and successfully acquiring much-needed grant money. It is a team-building exercise that has the potential to produce great rewards, while improving the efficiency of your organization and increasing the safety of your members.

Attend a conference

It does not take much to sign up for and attend a fire conference. On the surface, this may not appear to be much of a team-building exercise. Convincing a group of people to jump into a vehicle or get on a plane to go to a conference is not the exercise. It is what happens at these events that will provide your team with invaluable experiences.

Take FDIC for example. The annual Fire Department Instructors Conference (FDIC) is a major event for the fire service, bringing together tens of thousands of firefighters, instructors, and manufacturers to keep up with the latest developments, tactics, and equipment in the firefighting industry. The conference includes hands-on training (HOT), classroom sessions, and workshops. It also has award and remembrance ceremonies, vendor exhibits, and special events like the 9/11 memorial stair climb. These conferences, and ones like it, provide valuable opportunities to bond and learn together.

The passion you will hear and feel from the presenters and speakers will help you increase your commitment level (fig. 1–20). Having your team members at the event with you will spread that feeling throughout your organization. Ultimately, it will help your team take action steps that will help them function at the highest possible level. When you attend a seminar or conference on your own time, with your own money, you are making a commitment to better yourself. You are doing more than you have to do and more than anyone is asking of you. You will learn from the best instructors in the world, shake hands with icons in our industry, network with like-minded people, and come back with a first-class education in your chosen field. You and your team will be inspired to be better than you were before the conference started. That alone is worth the price of admission.

Fig. 1–20. Here is Bobby Halton addressing thousands of firefighters during the opening ceremony of FDIC in 2013.

We all had fun and learned a lot at my leadership development class at FDIC in 2014 (fig. 1–21). Each of us committed to enhancing our leadership presence within our organizations. I encouraged those in attendance to leave the event committed, not excited. Excitement is temporary. Commitment is following through on that thing you said you were going to do long after the mood you were in when you said it has passed. Attending a seminar or conference will present you with the opportunity to network with a group of highly motivated members of the fire service. Bringing team members with you to these events will provide you with an incredible burst of energy and enthusiasm that can improve your team's level of performance.

Fig. 1–21. To this day I still communicate with many of the people who attended my leadership development class at FDIC in 2014. We keep each other motivated.

Summary

These are not the only team-building exercise ideas you can incorporate to help develop a stronger, more united organization. You may have seen or used other techniques in the past that have worked well and produced desirable results. The point is, when you work to establish a goal, get people involved, and solve problems to improve your organization, you will build team spirit in the process. When you find a technique that works for your team, build off of it and continue to push yourselves until you achieve that extra degree of difference that often separates the good from the great.

References

1. John C. Maxwell and Steven R. Covey, *The 21 Irrefutable Laws of Leadership: Follow Them and People Will Follow You*, 10th Anniversary Edition (Nashville: Thomas Nelson, 2007).

2. Stefanos N. Kales, et al., "Emergency Duties and Deaths from Heart Disease among Firefighters in the United States," *New England Journal of Medicine* (Waltham, MA: Massachusetts Medical Society, 2007).

Preparing for Success

2

Preparation is often the difference between wanting to achieve success as a team and actually achieving it. The word *preparation* is often used when discussing team efforts, but the only way to determine if a team is truly prepared is by watching the way they perform when their skills are needed. It is no secret that the profession of firefighting is a difficult one. There is nothing routine about what we do. We cannot become complacent just because we think we are well prepared. Preparation is a never-ending process.

You have probably heard of the 80/20 rule, which states that 80% of our productivity comes from about 20% of our efforts. This rule relates to many areas of life. You could say that 80% of the work on your team comes from about 20% of the members, or that 20% of your tools will do 80% of the work. However you use it, the 80/20 rule does exist, and seems to work out in uncanny ways, even when it comes to preparation. For example, 20% of your skills will be used 80% of the time, but the skills you need during the remaining 20% will be more likely to make the difference between life and death. This is where preparation comes into play.

Preparation is the action of making ready or being made ready for use. A synonym for preparation is the word *training*. Training is the most important function in the fire service. Without training, we simply cannot be prepared. Textbooks that probationary firefighters receive while in the academy contain more than 1,000 pages of basic skills that every firefighter should possess. That's *basic* skills! Firefighters simply cannot prepare enough. Even when they get the basics down, they still have to prepare for low frequency incidents like terrorist attacks or a major natural disaster. No matter how hard we trained in the tristate area, no one expected what occurred on 9/11 or during Superstorm Sandy.

Training should be a lot of things, but it is not always enjoyable. While conducting drills we need to push ourselves, and hopefully we *want* to push ourselves. Failure to prepare properly is unacceptable, but failure itself is not always a bad thing. Failure during training exercises can provide you with a great advantage. When you fail, you figure out one more way that does not work. When you fail, you also are provided with the opportunity to further develop your ability to adapt. The goal should not be to train until you get it right, but to work past that moment and continue to train until you can no longer get it wrong.

If you or your team members ever foolishly begin to think you can slack off, remind yourselves that what you do not know can ultimately be the difference between life and death. What you do wearing your helmet is only as good as what you put in your mind. This chapter will help you prepare for the many challenges, routine and otherwise, you and your team will encounter.

It's not the will to win that matters—everyone has that. It's the will to prepare to win that matters.

—Paul "Bear" Bryant

Decisions Are Like Tattoos

I have a tattoo. I made the decision to get one when I was in my early 20s, and I am still not sure why. Maybe peer pressure influenced me, since everyone I knew had a tattoo. I do not regret it too much, mainly because it is not generally visible. I rarely think about it, but whenever I do, I usually wonder what I was thinking.

You may have a tattoo as well, perhaps some tribal art or a wild animal. Maybe it is a fire helmet or Maltese cross, or perhaps it is the name of your child or your spouse, or even your ex-spouse (ouch). Maybe you love your tattoo. Or maybe you are like me and wonder what you were thinking when you got it. Whether you love it or not, there is one cold, hard fact: once you put a tattoo on your body, you cannot change that decision. It is true that you may be able to cover it up with another one, or you may even be able to get it removed after 108 laser treatments. However, you will always have made that decision to get it in the first place.

This is not necessarily a bad thing. It is just something we all need to put into the proper perspective. I want you to start to look at decisions the same way you look at tattoos: they are permanent. You may be able to change the outcome, but you cannot go back in time and change the decision you made. You have to work with it, and if it is wrong, you will have to find a way to adjust. The key to success in life is to get good at making better decisions.

Preparation is one of the key components to good decision making. With education and experience, we have a better chance of making the right choices. Another great way to make better decisions is to discuss options with your team. When I have to make a big decision that I know will have an impact on my life and the lives of the people I love, I call people I respect to get their thoughts on the situation. On the fireground, however, I do not have that luxury, which is why we have to prepare beforehand.

The key to success in life is to get good at making better decisions.

Some incidents are unusual in nature and present problems that even the most seasoned fire service professionals rarely encounter. At incidents like these, I am not above using the collective judgments of those who are on scene with me. There have been times when I have been IC and had an officer come to me to discuss an unusual situation that was occurring. Together, we quickly created a game plan. However, once we made a choice and implemented a strategy, it was permanent. We could always revise our plan after it was implemented, but we could never go back and change the initial decision we had made.

To become a more effective team leader, it will be essential for you to understand the importance of decision making, both for yourself and your team. Preparing your team to make better decisions will help you step up your teamwork.

Activity vs. Productivity

Have you ever wondered how your team can be incredibly busy doing important activities but still not come close to achieving your desired results? To make matters worse, you notice that other teams seem to be knocking it out of the ball park. You know your team is working just as hard, maybe even harder than the others, but you just cannot seem to make the progress that you want and expect. Perhaps your entire team feels as if they are running up the down escalator. It does not make sense. You know you are putting in the sweat equity needed to succeed, so why is success eluding you?

The answer is simple. People fool themselves all the time into thinking they are busy, sometimes too busy to do any more than they are doing. Often it is true that they are busy, but they are busy doing the wrong activities. Have you ever seen the rock demonstration? If not, you might find this interesting.

Imagine you are sitting at your desk and you have a one gallon bucket in front of you. Also on the desk are four different materials: water, sand, pebbles, and large rocks. Each of those materials represents a different activity that is required for your team to be more productive. Using success on the fireground as an example, water would represent maintaining your tools and equipment. The sand represents self-education, which includes such things as watching training videos, attending seminars, and reading SOPs, articles, or books. The pebbles represent learning about the incident command structure and the specific roles of every assigned position and company. The bigger rocks represent hands-on training on the thing your team will be doing the most—firefighting.

All of these activities are extremely important, but most new officers fill up the buckets of their team members with water, sand, and pebbles, and leave no room for anything else. Their bucket is full before they get to the large rocks. In other words, they are so busy doing things like maintaining tools and reviewing documents and procedures that they leave no time for the hands-on training. They fail to make it a priority to practice the skills, procedures, and techniques they will need to utilize at the actual incident (fig. 2–1). No one is disputing the fact that all of the activities listed above are important. The illustration simply represents how most teams spend too much time on some activities and not enough on others.

Fig. 2–1. We must make time to practice the correct skills, procedures, and techniques that we will need to rely on when the incidents occur.
Courtesy: Sid Newby.

To correct this, choose the most important activities your team must do on a daily basis in order to achieve success and make those activities your top priority. Solving this problem requires that you identify the difference between *activity* and *productivity*. Begin to fill your bucket with the most productive success-producing activity first. Consider the fact that when you fill the bucket with water first, there is no more room for sand or pebbles, let alone larger rocks. Now, let's reverse the situation. If you put the rocks in first, there will be room for pebbles and sand to fill in between the voids. And whatever spaces the pebbles and sand do not fill, the water surely will.

Let me further clarify this concept with another example. Consider people in real estate sales whose goal it is to sell houses. If they fill their calendars with qualified showings (large rocks), they will find it easy to educate themselves on available inventory (pebbles) and make calls to prospects and team members throughout the day (sand). They then will be able to read educational material about how to become better salespersons (water) with any remaining time. The result will be more sales and increased income because they are emphasizing the activities that matter the most. If, on the other hand, they fill their calendars with less important activities and do not have time to show houses, they will not have profitable careers.

If you and your team are facing the common problem of being busy but not productive, it is time to be completely honest and revise your strategy. Revising a strategy is not a one-person activity. On the fireground, the team leader (IC) reviews, evaluates, and revises strategy and tactics based on the progress reports he or she receives from the other firefighters on the fireground. The IC may know what is happening in the command post and in front of the building, but may not necessarily know what is happening inside the structure. This information has to come from others.

Never mistake activity for achievement.

—*John Wooden*

Early in my career, as a probationary firefighter, I was operating a hoseline inside a building. My partner's air ran low, and he left me to go get another bottle. I was so new that I did not know I was supposed to leave the structure when he did. Since he said, "Wait here, I'll be right back," I thought that was the right thing to do. The incident happened so long ago, I do not even remember where our officer was. He may have left to change his bottle 5 minutes earlier for all I know. From a safety aspect, we were all very foolish, but I was doing what I thought was right at the time. I kept the nozzle open and swept the fire until the flames went out. A feeling of pride came over me because I thought I did something great. I did not have a radio at the time, but I slowly began to make my way back to the stairwell so I could find the IC and report that the fire was out. As I found the stairwell and began descending to the first floor, I heard the air horns on the fire engines blow four consecutive times. This is the signal to exit the building because we are changing from an offensive to a defensive strategy. I hurried down the stairs and exited the building. I was shocked when I turned around and saw fire coming out of every other window on the floor I was just on. I had no idea what was happening anywhere else on the fireground: I only knew what was happening in my own space. That mistake could have cost me my life.

As a tour commander, I do not ever want this type of experience to happen to a firefighter on my team. We train on all aspects of firefighting operations and communication precisely for that reason. What if a similar event happens to someone on your team, and you are not aware of it because nobody says anything about it? You cannot properly prepare and become more productive unless you are aware of what is happening on your team. The only way to be informed is to receive feedback from your team members. Ask your team what they think was effective. Then ask what they think they could do better. Remember, though, that feedback alone will not make your team more productive; it is what you and your team choose to do with that information that will move you from being an active team to a productive team.

After Action Review

One way of attaining information is with an after action review (AAR). An AAR is a structured review or debriefing method for analyzing what happened, why it happened, and how things can be done better. The formal AAR was originally developed by the US Army. The concept works and has been adapted by many nonmilitary organizations, both domestic and international. The fire service employs the concept of AAR but calls it the *post-incident analysis* (PIA). Both AARs and PIAs can be conducted after every incident simply by asking these four questions:

1. What was our overall mission?
2. What did we do well?
3. What could we have done better or differently?
4. Who needs to be informed?

One of the best ways for you to prepare your team is by gathering after every unique challenge, no matter how small, and conducting an AAR. This tool can be used to gather information and build accountability within your organization. This type of review can also be used to help a leader establish his or her intent and explain the desired end result, either formally or informally:

- Formal AARs and PIAs are conducted with every member on the team. They are intended to provide the maximum educational and training benefit.

- Informal AARs are usually conducted with fewer members and prior to the formal one. The main goal is to discuss what worked and take immediate corrective actions in case another incident occurs before a formal review.

Regardless of what you call it, and whether it is formal or informal, the main purpose is to discuss what went right and what went wrong. Identify and take note of actions that were correct, discuss what needs to be improved upon, and find the best way to perform a specific task or mitigate a specific type of incident. Then prepare the team so the next time you respond to the same type of call, you are better (fig. 2–2).

Fig. 2–2. After action reviews enable teams to take information and experiences acquired from one incident and use them to better prepare for future incidents.

If you are not discussing ways to improve, you are not leading a team. In fact, if you are not focused on finding ways to prepare and improve, you are failing in the role as a team leader. Continuous improvement is the goal of every high-performance organization.

Too many people are afraid to critique performances because they are afraid to raise some form of legal liability. They are afraid that they may have done something wrong, which contributed to a poor outcome. Let's get away from this mindset and start realizing that we have to analyze our performance if we want to improve. And when problems occur and mistakes are made, we cannot afford to take it lightly. We owe it to ourselves to look deeper into why the problem occurred and find its root cause.

Root Cause Analysis

Root cause analysis (RCA) is a method of problem solving that tries to identify the basic causes behind unsatisfactory results. The RCA concept has existed for decades in the business world. The concept is simple—look for the root cause or the reason why something is not being done correctly.

When it becomes apparent there is a problem that needs to be solved, it is important to get down to the bottom-line reason why the problem is occurring. Start by making a list of what went wrong. Let's use the following example of improper ventilation or failure to vent the roof, which contributed to the loss of a structure.

1. Begin by asking, "Why did this happen?" In this case, the answer is that we were not able to get the roof open.

2. Equipped with that information, again ask, "Why did this happen?" In this instance, the answer is that we were not able to get a crew on the roof early enough in the incident.

3. Finally, with the knowledge of that answer, again ask, "Why did this happen?" In this case, the answer is that the first-arriving ladder company had to perform a primary search. The second ladder company was coming from another town on mutual aid and took 15 minutes to get on the scene.

In that example, the root cause was not that you did not get the roof open; the root cause was that you did not have a ladder company available soon enough to perform that function. The solution would be to get that second company to the scene more quickly so you have a team available to perform that function. Now you can solve the problem.

The RCA concept is simple. Identify a problem and then ask, "Why did this happen?" until you get to the point where you identify the root cause (fig. 2–3). The process is equivalent to child who keeps asking, "Why?" "Why?" "Why?" However, keep in mind that there may be multiple root causes that led to the outcome.

Fig. 2–3. The RCA concept is simple. Keep asking *why* until you get to the bottom line.

A good fire service example of the importance of determining RCA is freelancing. Freelancing occurs when a person works outside of an established action plan. When a firefighter freelances, the initial reaction from almost everyone is that the individual is reckless. Research has shown us, however, that the top five causes of freelancing are the following (not in order):

1. A renegade firefighter
2. A weak company officer
3. Poor training
4. An unclear action plan
5. Poor communication

Four out of the top five reasons cannot be attributed to the firefighter, but instead to those at a higher level—the officers. But do not stop there. You have to keep asking *why* until you finally get to the root cause.

Do not look at a problem only on the surface and think you have it all figured out. Instead, look for the root cause. It is the only way you can be assured that you are taking the proper corrective actions and learning from mistakes of the past.

If you are not focused on finding ways to prepare and improve, you are failing in the role as a team leader.

How We Learn

When most organizations select new employees and team members, an individual's previous experience is a key factor. In the fire service, we do not always have the luxury of choosing our team members. In some organizations, civil service testing or some other predetermined ranking process does most of the decision making for us. At the very least, these systems limit our choices. Experience is rarely a major consideration with the hiring of new employees. Quite frankly, how could a person gain experience being a firefighter unless the person was actually doing the work that firefighters do? For this reason, only a comprehensive training program can prepare firefighters with the skills needed to perform the necessary duties.

Training must be continual, with repetitive drills. Proper firefighting is built on a foundation of perishable skills, and unused skills fade quickly. Continuous training is the only way for us to hone our skills and learn new ones (fig. 2–4). To meet the needs of fire service training requirements, a training officer (TO) is usually appointed to coordinate all training activities. In most cases, the TO will develop a 6- to 12-month training schedule. The schedule will consist of in-house drills and field drills. If the area of training requires an expert who is not currently a member of the team, outside instructors will be brought in to provide members with the necessary information. All drills are documented by category and include a list of what was taught, as well as a list of participants. This helps ensure that all team members satisfy the necessary training requirements set forth by the state and their department.

Fig. 2–4. Continuous training is the only way for us to hone our skills and learn new ones. *Courtesy:* Donald Colarusso.

The TO is a very important member of a team. That individual may not work on the apparatus or respond to calls, but a smart leader will develop and maintain a good working relationship with the person fulfilling that role. Whenever you determine the need for specific training, you can speak with the TO about your options. Whether you have a TO or not, your goal as a leader should be to learn and teach something new every day. In order to do this, it is important to understand how people retain information (fig. 2–5).

> **How We Learn and Retain Information**
>
> We Retain:
> - **10%** of what we **read**
> - **20%** of what we **hear**
> - **30%** of what we **see**
> - **50%** of what we **see and hear**
> - **70%** of what we **say as we talk**
> - **80%** of what we **experience personally**
> - **90%** of what we **say as we do things**
> - **95%** of what we **teach to others**

Fig. 2–5. How we learn

Looking at those percentages, it is easy to see why even a well-organized and clearly delivered lecture is not the most efficient way to learn new skills. When attending a lecture, most people hope to absorb the majority of the information being presented. It is not until later, when they need to recall and implement these new skills in real-life situations, that they realize they have not retained or understood as much as they hoped. Unfortunately, the instructor may not be readily available for guidance.

There are many factors that influence the effectiveness with which learners can absorb and apply new skills and knowledge. In the book *Management in the Fire Service*, authors Harry R. Carter and Erwin Rausch list the eight conditions that instructors should be conscious of when preparing a class. These conditions include: pace, absorption, understanding, practice, order, recognition/recall, new knowledge, and feedback.

Conditions that influence learning

1. Pace. All individuals have different capabilities of learning. Some people understand quickly, others have good memories, and still others have long attention spans. Most of us, however, have the attention span of a goldfish (a scientific fact, as discussed further in chapter 3). If the presenter's pace is too fast or too slow for the learners, he or she can lose them early in the presentation.

2. Absorption. A combination of presentation and hands-on training is the most effective way for our team members to absorb the material being presented. Taking the practical portion out of a class may save time but may defeat the purpose of the lesson.

3. Understanding. Skills that are learned (or practiced until they become habits) and thoroughly understood are retained better than skills that are learned only by memorization. The best way to see if the student comprehends the information is through testing. Keep in mind, however, there is a big difference between those who understand and those who have just memorized some facts.

4. Practice. Consistently practicing skills over long periods of time is more effective than compressing the same amount of practice into a single period. Repetitive, short sessions lead to better retention.

5. Order. The order in which the instructor presents new materials is just as important as the information itself. Teaching complex techniques before establishing an adequate foundation is a tactical error on the part of the instructor.

6. Recognition/recall. When testing to see that people understand a skill, emphasize recall of the skill, not just recognition of when the skill is needed. The point of learning in the first place is to train people to know *when* and *how* to do something.

7. New knowledge. The point of your class is to educate. Your goal should be to present new information, but when teaching people new skills, do not neglect to reinforce old ones on the same subject matter that are still relevant and important. Failure to do so could result in confusion.

8. Feedback. Without asking for feedback or including the learner in discussions, the instructor will not know if the information being taught is actually being absorbed.

Take those factors into consideration whenever you are preparing to instruct. The more time you devote to creating the proper educational environment, the better off you and your team will be in the long run. For more on presenting and public speaking, refer to chapter 3.

Identify your learning goals and objectives

Never provide training for your team without first identifying your goals and objectives. Analyze the needs of your team and contemplate the following four questions:

1. What does my team need to learn?
2. What type of learning experience will be most effective?
3. Who can best teach these skills?
4. How can progress be measured and ensured?

Once you answer those questions, decide the best time to provide this training and make arrangements to move forward. Do not hesitate to pass on essential information to your team. The difference between 211° and 212° could simply be a lack of training in just a few of the areas where your team needs to become more efficient.

Unused skills fade quickly.

Individual learning obstacles

Most people, especially those on an energetic team, will have a positive attitude toward learning. However, regardless of attitude, some people have learning disabilities, many of which have not been officially diagnosed. You may have been in training with someone you thought was slow or disinterested, when in reality, he or she had a learning challenge and was too embarrassed to tell others. Be aware of this possibility when you see one individual falling behind the rest of the team. You may be able to determine whether this is more than an attitude problem with good communication and understanding. Add counseling and special training arrangements, and you may be able to easily solve this challenge and help build the confidence of this team member.

Look for training opportunities

You do not have to look hard for training opportunities. They are all around you. Smart firefighters train continuously throughout the day. They train just after roll call by taking a few minutes to go over some of the equipment and tools on their apparatus. They train around the kitchen table by discussing various scenarios and procedures. They train on routine calls by taking a few minutes before they get back on the apparatus to discuss building construction features or other pertinent information.

There are so many ways you can take everyday moments and turn them into an educational experience with just a little effort. One of our members brought 200 feet of twine to work. Every time his company went out to inspect buildings or investigate alarm activations, he made good use of the 10 minutes or so that it took for the officer to get information for reports. This team member would stretch the twine from the apparatus to a specific location inside the building to ensure that the 200-foot hoseline would reach it if needed in the future. As a team leader, you should also constantly be setting the example for how your members should interact with customers. When your efforts far exceed expectations, your team will learn from watching you. This became evident to me one day when a parade passed by our firehouse.

Every summer, the Hudson County Peruvian Day parade begins in the neighboring community of Harrison and marches through Kearny. The parade route is more than an hour long and passes by our firehouse. This Peruvian cultural event consists of people who are proud of their heritage and who are also committed to preserving their history and traditions. During the parade, they wear colorful and elaborate outfits and dance as long as they can. It is incredibly entertaining and uplifting. On this particular day, however, the celebration seemed much less festive than usual, and the reason for this was obvious. It was 100°F, the hottest day of the year. Our firehouse was more than a mile from the starting point, and some of their outfits easily weighed as much as firefighters' turnout gear. This was a recipe for disaster.

I was up in my office as the parade began passing by. Our members were doing what they usually do, enjoying the parade and waving to people as they pass. I did not have a view of the parade from my office on the second floor, so I came down to watch for a few minutes. As

I left the air-conditioned living quarters and walked on the apparatus floor, I was reminded about how hot it actually was outside. Without breaking stride, I walked over to our garden hose, turned it on, put the nozzle in spray position, and positioned it to create a makeshift outdoor shower. The newest member of our team looked at me with a puzzled expression. Within moments, dancers from the parade broke away and danced under the water before returning to the parade line (fig. 2–6). One group after another followed suit. They all said thank you, but they did not need to; their expressions said it all. The young firefighter looked at me and said, "Good thinking, Chief."

"Don't ever forget, our job is to serve people and exceed their expectations," I replied, before returning to the office.

I learned just as much that day as the young firefighter did. He learned that we are always serving others. I learned that our team members are always watching and learning from the example we set.

Fig. 2–6. You do not have to look hard for opportunities to educate your team members on how to provide great customer service.

CHAPTER 2   *Preparing for Success*

Break it down

At large-scale incidents, every crew and individual has a specific task. When a team of firefighters is assigned as a rescue company, their job is specific. They will not be putting the fire out or venting the roof. They will be doing rescue-related activities. They are a small, but incredibly important, part of the overall big picture. However, the jobs that the other companies are doing at that same incident are also important (fig. 2–7).

Fig. 2–7. Every individual needs to know his or her job and understand how it fits into the overall picture.

If you look at the incident in its entirety, it can be quite overwhelming. In fact, the combined actions of every single company that responds to the scene can be perceived as overwhelming if you do not prepare properly. Instead of looking at the entire task, efficient firefighters train in a systematic way. They break their actions down into manageable steps and practice those steps until they become second nature.

During my senior year in high school, I signed up for a wrestling club that met two nights a week. I was an average wrestler but wanted to improve. The other members of the club read like a who's who of

New Jersey wrestling. Each had each achieved a high level of success in the sport. One of the athletes in the room was a year older than me. He was a three-time high school state finalist and two-time champion. At that time, he was in contention for an NCAA championship. I remember practicing with the other wrestlers while the former State Champion was off in a corner of the basement wrestling with his older brother, who was an All-American and a Division III national champion himself. These two brothers rewrote record books for their schools and were considered wrestling royalty in the state of New Jersey. What struck me as odd was what the younger brother was doing over in the corner. He was practicing a double-leg takedown, over and over. He was continuously repeating the motion, without much resistance from his brother, for nearly an hour. It made me tired just watching him. This guy was one the greatest wrestlers to take the mat, and what was he doing? He was rehearsing a simple move, over and over.

It made an impression on me. That night, I learned why this young athlete was one of the greatest takedown artists in the sport. When everyone else was at home resting, he was breaking down his match to the most basic moves, creating muscle memory, and making each move feel natural, and it paid off. Before the end of his storied career in college, he achieved NCAA Division 1 All-American Status twice, finishing third and second. He was also a three-time NCAA Division III National Champion and finished third on the Olympic ladder.

Repetition provides many benefits. One of the benefits is that it enables you to become so good at doing a task that you do not fall into the trap called *task fixation*. When you give all your attention to performing a single task, you tend to miss critical information in your surrounding environment. It is biologically impossible for the brain to multitask the act of paying attention, so the key to successfully (and safely) performing a task is to create muscle memory. The subconscious is the neural location of muscle memory and habits. And the subconscious brain can multitask.

Muscle memory

Muscle memory, also known as *motor learning,* is a form of procedural memory that involves consolidating a specific motor task into memory through repetition. When a movement is repeated over time, a long-term muscle memory is created for that task, eventually allowing it to be performed without conscious effort.

It made an impression on me. That night I learned why this young athlete was one of the greatest takedown artists in the sport. When everyone else was at home resting, he was breaking down his match to the most basic moves, creating muscle memory, and making each move feel natural, and it paid off. Before the end of his storied career in college, he achieved NCAA Division 1 All-American Status twice, finishing third and second. He was also a three-time NCAA Division III National Champion and finished third on the Olympic ladder.

Fig. 2–8. Train the way you want your team to perform in real-life situations. *Courtesy:* Cindy Rashkin.

Imagine you are away on business or a family vacation. It is 11:30 PM and you have a busy day ahead of you, so you are getting ready to turn in for the night. Just then a fire starts in another hotel room on your floor. You hear the alarms sounding. You open your door and discover that the hallway is heavily charged with smoke. Since your room is on the 20th floor, you close your door, place damp towels around the bottom, and wait for the fire department to come save you. On the television, you are watching exterior views of the fire through a live news feed. As the firefighters hurry toward the building, carrying a high-rise hose pack, a reporter runs alongside, places the microphone in front of them and says, "How are you going to rescue the people?"

The firefighter responds, "We don't know. We've never trained for this."

How are you feeling right about now?

In the fire service, we train regularly on techniques for rescuing both civilians and firefighters because there may come a time when we have to perform these tasks (fig. 2–9). What are the most important things you would like your team members to do automatically? A professional sales trainer would want a team to learn how to overcome objections from their prospective customers. An NFL offensive coordinator would want the quarterback to be able to quickly read the defensive alignment at the line of scrimmage and quickly call an audible when something unexpected is seen. As a fire chief, I want my ladder companies to systematically vent, enter, isolate, and search an occupied structure like it is second nature to them.

As a team leader, it is your responsibility to determine which skills are absolutely necessary for the survival of your team and to prepare your team members to work in pressure-filled environments. We do this through proper training, and it is one of the reasons why I love the concept of real-time drills. They help people create muscle memory and second-nature reactions.

What will happen if you and your team train and create muscle memory the way Olympic-caliber athletes do? You *will* become more efficient.

If your team trains the way Olympic-caliber athletes do, you will become more efficient.

CHAPTER 2 — *Preparing for Success*

Fig. 2–9. Practice the techniques you want your team to be able to perform automatically. *Courtesy:* P.J. Norwood.

Practice under pressure

Once you get the basics down, it is time to improve your game. Try to create the pressure you will feel during real-world situations. For firefighters, this can be done at fire academies or with live burn training. Do not try to create perfect conditions; you will want to practice with distractions going on around you because the fire scene presents an abundance of distraction. This is one of the main reasons why we develop mask confidence obstacle courses and train on self-rescue techniques. When we do this, we are forced to practice under pressure and overcome challenge after challenge. One of my favorite things to witness is the obvious increase in confidence in firefighters as they complete one of these courses. The same can be said for a group of firefighters who just found and rescued the mannequin and put out the fire at their local fire academy.

Pressure-filled training exercises benefit leaders as well. Chief officers and those who will be taking command at incidents need to train under pressure for their job just as much as line firefighters do. They need to learn how to perform in hostile environments. If they cannot handle pressure and let their nerves get the better of them on the fireground, the people under their charge have more to overcome than just the fire.

There is one caveat to training in high-pressure environments: do not throw every obstacle in the book at a newly developing team. You want to first give them a chance to learn the necessary skills. I have seen training officers and academy instructors who developed training scenarios that Navy Seals could not get through. Make sure the individuals you are training know the basics before you throw them into the lion's den.

Every day you come to work, expect a fire.
Every fire, expect that people will be trapped.
On days when there are no fires,
train for the days when there will be.

Preparation Leads to Confidence

Firefighters need to be taught to ask questions that begin with "what if," such as the following:

- What if the fire extends into the attic?
- What if the thermal imaging camera fails?
- What if the wind changes direction at a hazardous materials incident?

There is no guarantee that things are going to go as planned. How we prepare for adversity determines the outcome. I am sure you remember Captain Chesley B. "Sully" Sullenberger and his crew on US Airways Flight 1549. On January 15, 2009, Sully was flying a plane carrying 154 passengers as it took off from LaGuardia Airport. During takeoff, the plane hit a flock of birds and lost power in both engines. The plane rapidly lost altitude. Sully tried to return to LaGuardia, but he quickly realized they would be unable to reach an airport. According to the FAA transcript, when the air traffic controller announced that "Cactus 1549" had clearance to land at runway 1 in Teterboro, in New Jersey, Sully replied, "We can't do it."

"Okay, which runway would you like at Teterboro?" the controller asked.

"We're gonna be in the Hudson," Sully said, in a calm, composed manner.

"I'm sorry, say again, Cactus?" the controller replied.[1]

He had one option, to land the plane in the Hudson River. Sully and his crew did just that. All 154 passengers survived.

Was this really a "miracle on the Hudson"? Everyone who listens to the recording of Sully announcing that he is going down knows the definition of cool and confident under pressure. Where does a person gain that type of posture? The answer can be summed up in one word: training.

Pilots use a simulator called "the box," which is not a term of endearment. Even skilled pilots would admit there is nothing pleasurable about sitting in the box and being put through the wringer. Pilots don't practice landing in the Hudson River, but they do all practice a variety of emergency scenarios inside these sophisticated simulators.

Shouldn't the fire service have simulated training devices like pilots have? Fortunately, we do, and they are called fire academies. These academies exist so we can train in a controlled environment. Working fire exercises, propane tank fires, mask confidence courses, flashover simulators—are all examples of the type of training that is likely available to your team. Even if you do not have access to a fire academy, you can still plan elaborate training drills that will allow your team to hone their skills. Your team cannot afford to show up unprepared and hope for the best. The work has to be done beforehand.

*If I had eight hours to chop down a tree,
I'd spend six hours sharpening my ax.*

—Abraham Lincoln

Sweat More, Bleed Less

When I was considering taking a job as a firefighter, I was at my friend Michael McDowall's house, talking with his father. "Big Bill" was a well-known police officer in town, and he was an absolutely terrific guy. He always gave me great advice, so I spoke with him about my concerns regarding the job. He knew my father was a firefighter, and he knew it was a great profession. The advice he gave me was simple and honest, "It's a great job and you'll do well, but be careful because jobs like ours have a tendency to make some people lazy."

At the time I did not fully understand his advice. How could a firefighter or law enforcement professional be considered lazy? After several years on the job, I started to understand. Career firefighters are paid to be ready to respond quickly; that is why we are on duty at the station 24/7. A young, aggressive department with a solid leader will utilize that time to its fullest and find the right balance of training and downtime. But what if you are a young firefighter on a shift with a bunch of veterans and an officer who is at the tail end of his or her career? What if that officer is more concerned with his personal needs than he is with your needs or those of the team? You may find yourself in a scenario where your team is not spending nearly as much time preparing as they should be.

When this happens, a young firefighter has two choices—conform or step up. A maxim of the fire service is that the more you sweat on the training ground, the less you will bleed on the fireground. I encourage you to take that one to heart.

The more you sweat on the training ground, the less you will bleed on the fireground.

—Unknown

If you are on a team and you see signs of laziness and complacency, say something. Better yet, do something. You do not need permission to attend educational seminars or read books (fig. 2–10). On that note, I sincerely congratulate you for reading this book. The fact that you purchased and are reading this book says you are different from most. You do not just want to exist; you want to be part of a winning team. More than that, you want to help develop a winning team. That is one of the most important qualities you will need from the other members of your team—willingness. Now it is time to prepare the team.

Fig. 2–10. You do not need permission to attend an educational seminar or read books.

Daily Method of Operation

Every individual and team member has a daily method of operation (DMO). Even teams that fail have DMOs. The difference between teams that succeed and teams that fail lies largely in their DMOs. The fact is, teams that fail do not think about what they are doing daily and how it affects their overall performance. Their bad habits are on autopilot. Someone once defined the word *insanity* as doing the same things over and over but expecting different results. Analyze your DMO, and you will be able to determine if you fall into that category.

In order to create team synergy, it is essential that your days are built upon a solid foundation. Thinking positively, acting productively, and making decisions from a focused, centered, and purposeful place are all components of a solid foundation. From there, you will have to determine what actions you need to take, as an individual and as a team, to get the results you desire. This is easier than you may think. It begins with being honest about the results you are seeing. This is where you would benefit by asking yourself those three questions I listed in the culture development exercise in chapter 1:

1. What are we doing well that we need to keep doing?
2. What are we doing incorrectly that we need to stop doing?
3. What are we not doing at all that we need to start doing?

You know which activities are producing good results and which are not. These questions will help you identify where you are succeeding, where you are failing, and what you need to change. By asking these three questions, you can separate what works from what does not work and begin to develop a more effective DMO. Once you do that, you should create a DMO creed. It can be as simple as the one below, which my team follows.

Our Daily Method of Operation:

- Every day we train for a minimum of 3 hours.
- Every day we adhere to all departmental rules, regulations, and procedures.
- Every day we interact with others in a courteous and respectful manner.
- Every day we provide service to others, and we exceed their expectations.
- Every day we work on developing other team members as well as ourselves.
- Every day we work to achieve the desired goals of our team and organization.

If you intend to lead others, you need to have a DMO for yourself as well. You can create one for your personal and professional life. For example, the DMO I follow as an author and speaker is the following:

- Every day I read to educate myself.
- Every day I write to help and educate others.
- Every day I ask questions and listen.

- Every day I teach from the heart.
- Every day I solve problems.
- Every day I give thanks.

That last one is important. Motivational speaker Les Brown said, "Any day I wake up and there's not a white chalk outline surrounding my body is a great day." I happen to agree. I give thanks for my family, our health, and another day.

When you have a DMO, over time, you will understand the power of the *compound effect*. Small actions, repeated daily, produce the results you desire over time. You may also want to make daily checklists. This is one of the methods I use to stay focused on what we need to accomplish as a team. It is simple. All you have to do is take a piece of paper, write down the things you need to accomplish during the day, and put a check box in front of it. As you accomplish the task, check the box.

Insanity is doing the same things over and over but expecting different results.

—Unknown

Eat the Ugly Frog First

Brian Tracy has written a book entitled *Eat That Frog!: 21 Ways to Stop Procrastinating and Get More Done in Less Time*. The title references Mark Twain's theory that if you eat a frog first thing in the morning, chances are it will be the worst thing you have to do all day, so everything else should pale in comparison.

Tracy's theory is that if you have two difficult tasks to accomplish, you should take on the more difficult, less desirable one first. In other words, if you have to eat two frogs, make sure you eat the ugly one first.

When you get the worst item off your plate early on, this will enable you to get that task out of your mind, allowing for a more productive day. It is all about "state of mind" and the need to challenge your team right out of the gate in the morning. Get the most difficult task accomplished, and it will clear your mind and free your day so you can be productive without the weight of that task hanging over your head.

I encourage you to introduce this concept to your team. It will make a big difference in your overall productivity. You would also be wise to understand the importance of listening to others who believe they have discovered a better way of doing a specific task. What may be hard to you may not be hard to another person, so keep an open mind. Sometimes, with a slightly different approach, the ugly frog is not so ugly after all.

Never Settle

As much as I want to tell you that you can prepare for every situation, I know it simply is not possible in our profession. Too many things can happen. Too many scenarios can occur. Sometimes, they are a bit comical, like the call we had for my captain's three-year-old son, who climbed into large ceramic pot and became trapped. We arrived to find him kneeling in the pot, stuck at the waist, smiling from ear to ear because his daddy arrived to save the day.

Some incidents are not so funny. We once had an emotionally disturbed woman come to our dispatch center and smash out all the windows with an axe. Our dispatcher could be heard over the intercom screaming for help. One firefighter ran out of the station to the dispatch center in the rear yard to wrestle the axe away from her. Okay, in retrospect, it was a bit funny, but only because our cool, calm, collected dispatcher was suddenly screaming like a banshee.

And then there are those incidents that just do not make any sense at all. Like the time I was in my office writing reports when I heard the following radio transmissions from one of our companies that was driving around town.

"Squad 2 to dispatch. Can you put in an alarm for a llama rescue?"

"A what?"

"A llama rescue."

"Did you say llama, l-l-a-m-a?"

"Yes, as in South American llama."

Keep in mind, Kearny is in Hudson County, which is only a few miles outside of Manhattan. Hudson County has a population of 652,302, and Kearny is on a peninsula. For a llama to get lost in Kearny, you would think it would have to be beamed down from another planet (fig. 2–11).

Fig. 2–11. Captain Kevin Donnelly and the Kearny llama

There is no way to prepare completely for every scenario or all the possible things that can go wrong during a four-alarm fire or a terrorist attack. There are too many possibilities and too many variables. There is also little room for error. People are running away, and you are running into uncertainty. Sometimes you feel like running the other way yourself. However, if you have prepared properly, and you refuse to settle for the status quo, your training will bring you through. Even though your team cannot prepare for everything, remember that you must *never stop preparing*.

On a side note, the Kearny llama has become part of Kearny lore. He even has a Facebook profile. The last time I checked, we shared 94 mutual friends.

The Best Ideas Have to Win

One evening, sometime around 1990, a young couple in their early 20s was returning home after a night out on the town. The young man behind the wheel had had a few drinks earlier in the evening and was driving through town much faster than he should have been. As they sped down a hill toward the crossroad at the end of the street, the man lost control of the vehicle. It spun sideways, counterclockwise, stopping abruptly when it hit a utility pole. The car hit the pole with such force that the vehicle wrapped around the pole in the shape of a horseshoe. The exact point of contact was the passenger side door, and the woman's femur broke in half.

Another man witnessed the accident and called the fire department. Our members responded and upon arrival immediately began the extrication process. Kearny firefighters are no strangers to vehicle entrapments. When you have 24 bridges, two spans of the New Jersey Turnpike, and Routes 280, 1, 9, and the first elevated highway in the country running through your community, you get good at extrications. In fact, they are the most common rescue functions our firefighters respond to. They are very good at it, and they perform with speed and precision.

This entrapment was different. The man was able to exit the vehicle. Upon arrival, he was sitting on the curb, dazed and confused. The woman was enclosed in the steel wreckage. She was screaming, and rightfully so; her femur was protruding from the exposed flesh below her skirt.

Our members worked on the vehicle for 15 minutes, but they were unable to make progress. At the time, our tools were used and worn. We had been having problems during training evolutions with the cutters, but budgeting challenges did not allow for us to repair or replace them. The IC made the decision to call for the response of a rescue company from a neighboring department. They arrived within a few minutes, with newer, more powerful tools. After another 15 minutes, however, the firefighters were no closer to extricating the woman than when they arrived. To make matters worse, she was beginning to show signs of shock.

The firefighters were baffled. They could not seem to make any progress on the mangled scrap of metal that hardly resembled a car anymore. The group came together to discuss their options when one firefighter said, "If we could just take out that pole and work on the car from the other side, we would be able to get her out."

This was not an option. They could not cut the pole down because of the live power lines on top. Another firefighter paused. He looked at the vehicle then back at the group of firefighters. "There are enough of us here to pick up the car and move it away from the pole," he proposed.

It seemed like the best option. They collectively agreed to give it a shot. Within seconds, the firefighters were able to move the vehicle far enough away from the pole that they could access the passenger side. It took only a few minutes after that to remove enough of the vehicle to extricate the woman.

Of course, vehicles should be stabilized during extrication, and moving a car with an injured person inside is not the method you should choose when you have other options available, which is why I want to leave a disclaimer and encourage you not to do what the firefighters in this story did; however, this was a unique situation due to the condition of the tools, vehicle, and victim. In this rare instance, that was the best option the firefighters had available to them. That story demonstrates the power of a team working together and illustrates an important point: the best ideas have to win, even if they are not your ideas. A team leader has to be humble enough to realize that. Leaders should feel secure enough in their positions to allow other people to propose ideas and be smart enough to implement the best ones.

The best ideas have to win,

even if they are not your ideas.

What's Your Brand?

My team once gathered together for a post-incident analysis and I started off by saying, "We need a motto." A few of them looked perplexed. For years we had joked about other groups that had mottoes, but I think we all secretly wanted one that helped define us. The team had just fought an incredibly difficult fire. During the fire, one officer informed me that fragments of the drop ceiling were coming down on them, and they were backing out until the roof was opened. I made an announcement to all companies, directing them to work from safe areas so they could make a quick egress if necessary. The team had been inside the two-story residential dwelling for more than 15 minutes. Fire was blasting out the rear windows on the second floor, and yet they could not get water on it. At the time, I could not understand why.

When the crew from the engine company reached the top of the stairwell, they began making their way down the smoke-charged hallway toward the rear. They reached a bed, which was up against a wall. They could feel the intense heat and hear the crackling of the fire, but they could not see anything, not even the slightest glow. This was odd because it appeared that they were at the back of the building, and fire was blowing out the rear windows when we arrived. The captain told the nozzle man to open the line, but it bounced off the wall and steamed the room. They aimed the stream into the tiles of the drop ceiling, but were still unable to make progress. They backed out, made their way into another bedroom and tried again to reach the back of the house, where the fire was. Again, they were unsuccessful in locating the fire.

As it turned out, the headboard for the mattress at the end of the hallway was up against the door leading into the fire room. The room was used for storage and the family rarely used it. Because of the smoke, it was impossible for us to know it was a door behind the mattress and not a wall. I was watching the smoke push from the front windows, counting the seconds until I called them out. Suddenly, the roof was opened and the smoke lifted. The fire burned just enough of the door away for the firefighters to see a hint of flame through its cracks. They pushed the bed aside, busted open the door, and knocked the fire down.

They displayed true courage by staying in there and an incredible commitment to their mission and each other. As we were cleaning up at the scene, one firefighter approached me and asked, "Chief, I don't want to step out of line, but would you be okay if I offered free pizzas to the family? I've seen them in our restaurant before."

This particular firefighter helped run his father-in-law's pizzeria when he was off duty. He is a solid firefighter, with a strong work ethic and a heart of gold. Of course I was okay with him making the offer. The family of seven was displaced. I had just put them in contact with the Red Cross, and they were trying to figure out where they were going to stay. A free meal would be an absolute blessing for them. This firefighter, being the compassionate person that he is, spoke with them and told me the next day they came by and he treated them to a few free pizzas. He showed the true definition of the word *courtesy*.

"We need a motto," I repeated at the post-incident analysis. Microsoft is a brand, Coca-Cola is a brand, the United States Fire Administration is a brand, but what was our team's brand? "Think about it and let me know what you think," I said, before proceeding with our analysis.

We talked about the job and what we did right. We also talked about what we could do better. Before we finished, my cell phone vibrated. I looked down to see one of the firefighters in the room, Michael Richardson, had sent me a text message. It read: "Courtesy, Courage, and Commitment."

At that moment, the brand for the Kearny Fire Department, Group C was settled on. Our motto, Courtesy, Courage, and Commitment, embodies the following:

- **Courtesy.** Consideration, cooperation, and generosity in providing service.

- **Courage.** Mental or moral strength to venture, persevere, and withstand danger, fear, or difficulty.

- **Commitment.** An act of committing to a charge or trust.

We are a family, and it is not always easy. Yes we care about each other, but sometimes we also argue or disagree with each other like a family sometimes does. When push comes to shove, however, we protect each other like a family too. We are a family built on courtesy, courage, and commitment. If you intend for your team to achieve true success, you have to become a family as well (fig. 2–12).

Fig. 2–12. Firefighters have two families: the one at home and the one at the fire station. *Courtesy:* Tony Perez.

Marketing Your Company

Once you have a brand, your next goal is to create brand recognition. You want people to associate your team members with that brand. When they think of your organization, what comes to mind? Work on connecting your organization name with your brand. In Kearny, our goal is to be known for customer service, and you should absolutely want to be known for the same. We knew that to do this, we would have to continue working hard at fighting fires, but not stop there. A fire department's goal is to spend time fostering relationships with their customers—the taxpayers and citizens. We are in a service-based industry, and as with any other service-based organization, the customer has to be a priority. If you think the customer does not come first, think again.

Too many companies sell their product and then forget about the customer. They train their people how to sell instead of training them to connect with people and exceed their customers' expectations. They should be teaching their team members to be authentic, to connect, and to make a positive and lasting impression (fig. 2–13).

Step Up Your Teamwork

Fig. 2–13. Teach your team members how to connect with your customers.

Make service and professionalism your brand. Remember, branding is not simply a marketing tool. Authenticity is vital. If you do not create your brand with specific intent, you will be branded by others based on the experience they had with you. In the day where social media is king, this could be devastating for a team that has not clearly identified themselves.

One of the best qualities of the fire service is how various organizations respect each other. A firefighter from New Jersey can meet a firefighter from California and there will be an immediate mutual respect between the two. In the public's eyes, firefighters are a tribe. Everyone has a little "Monday morning quarterback" in them, but for the most part, fire departments rarely bash one another. Acting like an expert at fighting the fires you did not fight is a poor way to spend your time and energy. Learning from other people's experiences is important, but you should avoid criticizing what others are doing wrong as if you have never made a mistake. You may have a problem with another person in your profession, but do not lower yourself to a level you will regret. Put your energy elsewhere. Work on developing and strengthening your tribe.

Get the best minds on your team together to come up with creative ways to market your company. You will never be the best at anything if you do not first decide what you want to be known for. Once you know that, you can begin looking for the right people with the right qualities to complete your team.

Ten Qualities of an Outstanding Team Player

One of the key components of developing a winning team is to seek out individuals who have outstanding team-player qualities. Every organization should be committed to hiring quality people because 80% of the success your organization achieves will be a result of the people you bring in the front door. Skill sets will vary, but many team building experts agree that people representing their organization should at the very least possess the following 10 qualities.

Every organization should be committed to hiring quality people because 80% of the success your organization achieves will be a result of the people you bring in the front door.

1. Integrity

Without integrity, individuals and organizations can never become truly successful. Integrity is a foundational trait. You can build a winning team upon it, and you cannot build one without it. Integrity is a key component to building trust, and without trust, you have nothing. People define *integrity* in different ways. One of my favorite definitions is, "Integrity is doing the right thing, even when no one is watching."

Rick Warren, author of *The Purpose-Driven Life*, has another great way of defining this quality. He says, "Integrity is built by defeating the temptation to be dishonest." When an individual's actions and determination align with strong morals and values, and that individual is committed to the organization, you have found a person worth keeping around.

Integrity is a foundational trait. You can build a winning team upon it, and you cannot build one without it.

2. Committed

When a person is committed to the vision, mission, values, and ideals of an organization, he or she is exhibiting an essential quality you will want in a team player. Commitment breeds determination, and determination breeds perseverance. These are powerful traits that cannot be overlooked. Committed people keep their eyes on the prize, not the price they will have to pay. They understand there will be challenges and obstacles, but they look at them as minor bumps in the road. Committed people never take their eyes off the goal. They have adopted the mission of the organization as their own. When you fill a team with this type of people, you develop a strong, durable team that is ready to take on any challenge (fig. 2–14).

3. Enthusiastic

An enthusiastic individual with passion and few skills will always outperform an individual with great skills and no passion. Enthusiasm is exhibited by a sincere interest and exuberance in the performance of duties. People who are enthusiastic are optimistic, cheerful, and willing to accept challenges. They are eager to take on more responsibility and do not shy away from work. Instead, they seek better ways to do things. Every organization and team needs enthusiastic people because enthusiasm is contagious.

Fig. 2–14. Committed people are ready to take on any challenge. *Source:* John Harrison.

4. Skillful

Good leaders have learned to utilize the talents, skills, and abilities (TSA) of their team members. Great leaders look for individuals who possess the skills that are needed in order to add depth to their team. When individuals bring unique skills to the table, your team will begin to develop depth and a stronger union. When you can count on people to be proficient and deliver high-quality results in a specific area (or areas), you can improve the team by letting them "run in their lane." A team of people skilled in various areas enables a leader to allow each individual to work in his or her area of expertise.

5. Creative

In any arena, especially the fire service, we are constantly expected to solve unique problems. We often deal with situations we have only read about in books, so it is essential that we find creative individuals who have the ability to think outside the box. Your team will never suffer from too much creativity, but you can suffer from a lack of it. This quality cannot be stressed enough for those of us in service-based organizations. People who are constantly thinking about, training on, and trying new and better ways to do things can become great assets to a team that is dedicated to finding the "one best way" to accomplish tasks and satisfy customer needs.

6. Teachable

As important as it is to bring TSAs to the table, it is equally important to bring a desire to learn new skills. Outstanding team players have a thirst for knowledge. They read books, watch videos, attend seminars, and ask questions. They understand that they will not learn anything new by talking. If they are going to learn anything at all, it will be by asking the right people the right questions and by listening. They understand the importance of staying ahead of the competition. They know there is always a better way, and they recognize the fact that complacency is one of the biggest reasons why teams fail (fig. 2–15).

Fig. 2–15. Team players understand the importance of training. They are teachable and willing to learn new skills. *Source:* P.J. Norwood.

7. Sense of humor

As Benjamin Franklin noted, the only things certain in life are death and taxes. We are all on this earth for a reason, and stressing out over things we have no control over is not it. Firefighting is routinely listed as one of the most stressful jobs in America. On the other hand, it is also always listed in the top five for career satisfaction. One of the reasons for this is that most firefighters know how to take time out for a good laugh. Laughter is a powerful antidote to stress, pain, and conflict. Nothing works faster or more dependably to bring your mind and body back into balance than a good laugh. Humor lightens your burdens, inspires hopes, connects you to others, and keeps you grounded, focused, and alert.

8. Reliable

Whether a person routinely works in hostile environments under extreme circumstances or spends most of the day at a desk crunching numbers, reliability is an essential quality. Being reliable means your community, your customers, and your team members can count on you to get the job done, every time. Reliable people are capable of performing in routine circumstances, as well as in hostile or unexpected circumstances. Reliable people do not just start projects. They follow through and complete them.

9. Generous

Teams fail when one person wants to take all the credit. A generous team member understands the importance of distributing credit where credit is due. When every member of the team is working hard, and pulling an equal amount of weight, one person receiving more praise than the others will cause resentment and dissention. A strong leader will ensure that all members receive their fair share of the glory but will also look for members who are confident in their ability and unselfishly supportive of others around them.

10. Willing to work

All the qualities listed above are important, but without a solid work ethic, they are futile. A strong team consists of people who are ready and willing to get their hands dirty. You have heard the saying, "Many hands make for light work." That should be adapted to, "Many *working* hands make for light work." People who express a desire to take on new projects should not be taken for granted. When you find these individuals, welcome them with open arms and point them in the right direction. Mentor them and prepare them to mentor others down the line.

Determining who you want on your team is not a difficult process. Sometimes you can make that determination by asking one simple question, "Who would you want in your foxhole?"

Mentorship Programs and Succession Planning

When I first became a firefighter, I am not even sure the concept of mentorship programs existed. If they did, there were not many fire department heads who embraced the idea. I love the concept myself and think there is a need for mentorship programs at every level within any organizational structure. Even if your organization does not have a formal program, you can implement some form of a mentorship program on your team.

What would happen if you had an important presentation to give and you woke up feeling too sick to do it? This situation happens all the time. In order to fully understand and truly step up your teamwork, you must take time to develop others who can step up in your absence and do the job at hand.

One of the best sports-related examples I can remember of this principle was demonstrated by the 1990 New York Giants. Although starting quarterback Phil Simms had been the Super Bowl MVP four years earlier, Coach Bill Parcels believed in giving backup quarterback Jeff Hostetler playing time as often as possible.

In games where they built up a lead, Simms would play the first three quarters and Hostetler would complete the game. Some critics did not like this and thought Coach Parcells should play his starting quarterback throughout the entire game. It was interesting to watch Coach Parcells refuse to give in to the pressure. As they approached the playoffs, it was clear that Simms was ready for a postseason run. Fate had other plans. In a game against the Buffalo Bills, Simms was sidelined with a broken foot. Backup quarterback Jeff Hostetler would have to lead the team through the postseason. The newly assigned team leader was not only ready, he was determined. Hostetler led the Giants to a Super Bowl victory, making 1990 one of the most successful and memorable seasons in the franchise's history.

What would have happened if Coach Parcells had not given Hostetler the opportunity to develop his leadership skills on the gridiron? No one knows for sure, but there is a good chance the Giants' season would have ended the moment Simms was carried off the field.

The point is simple. Leaders must provide opportunities to let their team members develop new skills. True team leaders are dedicated to helping others reach their potential. Whether through a formal or informal mentorship program, the fact is, you must develop your team members.

True team leaders are dedicated to helping others reach their potential.

Mentorship programs focus on helping individuals and teams reach their potential. Succession planning prepares individuals (and possibly entire teams) to be able to seamlessly transition into another position. Succession planning is more than finding a new fire officer when the current one retires. True succession planning is an ongoing process that is designed to ensure the continued effectiveness of an organization at all levels. To be successful, mentorship programs and the succession planning process ultimately require a cultural shift in an organization. When the need is present, both concepts should be welcomed and implemented on a team.

Throughout the years, I have written a few articles on the topic of mentorship. It is also one of the topics I cover thoroughly in *Step Up and Lead*. You will not have a difficult time finding ideas from credible sources on how to begin a mentorship or succession program, but

I would like to stress the importance of choosing the right people to develop. Of course, we want to develop all of our team members, but there will be times when you have to choose a few key individuals to train for a specific position or job. Following is a list of eight things to look for when choosing the right mentee or successor:

1. Eager to grow/change
2. Willing to invest time
3. Positive attitude
4. Respectful toward you
5. Purposeful
6. Confident
7. Loyal
8. Accountable

Remember that these people are your ladder to success. Investing in them will provide a win-win scenario, benefiting you, your team, and your cause as much as it does them.

I was never assigned a mentor. In fact, not everyone believes in mentorship programs.

Jack Welch was the chairman and CEO of General Electric Company from 1981 to 2001. Welch never had a mentor. He thought the concept of a mentor was dangerous because it might make people think they need to become like their mentor instead of being their own unique selves. When asked to explain how a person could develop himself without being assigned a mentor, he said, "As a young person you want to have a nose for knowledge. You want to be looking at 10 people (and saying) I like the way this person speaks, I like the way that person makes presentations, I like the way that person leads people. And then you have to read voraciously because there is a tip everywhere. Everyone is a teacher."[2]

Welch cautioned about the mistake of trying to become someone you are not. He explained, "Don't try to be someone else. You want to make yourself an amalgamation of the best ideas you can put together with your personality and your style. Be you, but take some of these techniques you've seen from everywhere. But the most important thing is for you to get comfortable in your own shoes."

By the way, although the definition seemed obvious, I had to look up the word *amalgamation*, which essentially means "merger." If you had to do the same, welcome to the world of limited vocabulary. I feel your pain. However, I never let a limited vocabulary stop me from writing books (six to date). Do not let your limitations stop you from doing what you want to do either.

I was never assigned a mentor, but I secretly had plenty of them. I liked the way my brother Joe (also a deputy chief) took control at an incident (fig. 2–16). I liked the way Deputy Chief Mike Terpak from Jersey City FD and Anthony Avillo from North Hudson Fire and Rescue seemed to have an endless wealth of knowledge and an incredible way of communicating to a room full of firefighters. I worked with my brother, but the other two did not even work in my town.

Fig. 2–16. My brother, Deputy Chief Joe Viscuso. He is my best friend and my fire service mentor.

Even so, I have worked with plenty of positive examples. For example, I liked the way Gary Dye was calm under pressure. I liked the way Robert Osborn pushed hard inside the fire building. I liked the way Tom McDermott knew every detail about every building within his district. I liked the way Joe Mastandrea communicated with the public and represented the image of a firefighter. I can go on, but the point is, you do not have to be assigned a mentor. You can learn something not only from the people around you, but also from outstanding individuals whose work and character you admire, even those who work for other organizations and in different professions.

An Uneducated Firefighter Is a Dangerous Firefighter

We all need to do a better job of transferring knowledge to the younger generation. Too much is at stake. None of us want to see probationary firefighters learn about the dangers of truss roof or lightweight construction, or the signs of backdraft, through firsthand experience. We would rather they learn by hands-on training, watching videos, reading books, asking questions, and listening.

In our profession, young firefighters are great, but they are also dangerous because they lack experience. Everyone on the team is responsible for setting the right example and expectations. If they do not, they will encourage what I call a *coma of complacency*, which is an organization's worst nightmare. Establishing the right atmosphere is an incredibly important role for organizational leaders. Most jobs do not begin in high-pressure situations on the first day. However, in the fire service this very well may be the case. Mentors and team leaders need to make it their goal to teach something new every day. To do this, they have to also learn something new each day and lead by example.

If you are young and newly appointed into a leadership position, hit the books and watch the videos, attend the seminars, and begin networking with other people in similar positions. Social media has changed the game for those of us who are thirsty for knowledge. If you are seeking to gain knowledge in a particular area, you can easily connect with an expert in that area and begin communicating with them. You can also choose from thousands of free videos on the Internet to put together drills for your team. I once did a drill on "close calls." All I did was pull videos from YouTube, project them onto a screen, and have an open discussion about what went wrong and how we could prepare for and avoid similar circumstances. Try it yourself. Search for tips on areas where your team needs additional training. When you find something of value, share it with the team; discuss and learn.

People say experience is the best teacher, but in the fire service I have learned that experience can be very costly. Experience is not the best teacher; other people's experience is the best teacher. We must work diligently to educate ourselves. As a firefighter and officer, at the time of this writing, I have attended more than 170 classes at fire academies and other educational facilities. These were classes outside of our daily training routine at the firehouse. While attending these classes, I cannot even tell you how many times I would see the same individual sitting in one of the front three rows. He was a firefighter from a neighboring town named Brian Mulligan. I knew of Brian because I worked with his father in Kearny. One day I recall turning to a coworker who had attended a class with me and saying, "You see that guy? He's going to be a chief officer one day." It seemed obvious to me because of his eagerness and commitment to personal development. Sure enough, as I write these words, Brian has become a deputy chief for the Clifton Fire Department. I have a great deal of admiration for Brian and the Clifton Fire Department, a terrific organization. I share the story to confirm the statement that leaders become apparent long before they earn a title.

How does this relate to the development of successors? Simply put, those at the top of the food chain have to start looking at the field to see who is hungry to advance his or her career. Once you identify those individuals, spend time helping them develop. Your organization may work off of a promotional list of some sort. Maybe you use some type of performance or voting system. Whatever the process, you probably have a good idea who is next in line. Begin to identify how many team members you will lose through retirements during the next five years. Then identify who would be the likely candidates to fill those vacancies and train them for the positions. And do not miss the opportunity to have all members who are leaving your organization pass on their knowledge to the younger generation.

One organization told me they have retirees pass on their knowledge by completing a document where they list a minimum of five things they learned throughout their career that they want to share with new team members. The organization took all those documents and placed them in one book so future generations can learn from other people's experiences.

Experience is not the best teacher; other people's experience is the best teacher.

One of the biggest flaws I have discovered is that most organizations do not spend enough time preparing people for the next level. Every time I give a leadership development seminar, I ask the room, "How many of you have been through a formal training program for your current position?" Less than 20% of the people in the room raise their hands. This is something that needs to change. How can your team expect to reach its full potential if every leader has to learn by trial and error, instead of building on the foundation of those who served before?

Sharing information and helping develop strong team players is important in your quest to step up your teamwork.

References

1. ABC News, "FAA Releases Transcript from Hudson River Landing" (February 9, 2009), http://abcnews.go.com/Travel/story?id=6802512.

2. "Mentoring," C-SPAN interview (February 27, 2013), http://www.c-span.org/video/?c4369645/jack-welch.

3

LEADING TEAMS

Although this book is about team building, I would be remiss to not include this section on leadership. Much has already been written about leadership, and my book *Step Up and Lead* provides detailed information on leadership skills and traits. I will not repeat that information here, since *Step Up and Lead* and *Step Up Your Teamwork* are meant to be companion volumes. Together they can provide leaders with a strong foundation of knowledge to help them learn, grow, and improve. This book, in particular, is designed to be a follow-up book that teaches the person with adequate to above-average leadership skills how to assemble a group of people and help them achieve peak performance.

What's Your Title?

Are you a CEO, manager, supervisor, team captain, or project coordinator? Maybe you are a senior firefighter, company officer, battalion chief, or fire director, but regardless of your title, your primary job is to lead a team. It may be a team of 3 or 3,000. A strong leader knows his or her job is to take a group of people, align them with a common goal, get them to work as a team, and march them toward that goal. The bottom line is that you have to get the team to work together. To do this, you have to learn how to motivate, not manage.

What is your current mission? Are you trying to improve the performance of your team in high-pressure situations? Perhaps you are trying to continue to provide the same level of service you always have, only this time it is within the constraints of a significant budget cut. Whatever your mission is, getting others to understand the mission and work together is the key.

Teamwork is the ability to work as a group toward a common vision, even when events occur to obscure that vision. As firefighters, we fully understand what it is like to work toward a goal and push forward without seeing the environment around us (fig. 3–1). We have all been there, but a strong leader will be able to lead others through this type of environment in both hostile and nonhostile working conditions.

Fig. 3–1. Firefighters routinely enter structures and perform without the advantage of visibility. When your team members lose focus on their goals, a strong leader will help reconnect them to their vision.

Chapter 3 — Leading Teams

There are different ways to motivate and lead a team. Every leader has a distinct, personal style. When you think about great leaders, a specific individual may come to mind. I personally admire John Wooden and Zig Ziglar. Both will go down in history as great motivational leaders. Perhaps the thing I admire most about both of these men, however, is how they chose to live their lives. They were both godly men with strong values and a deep-rooted desire to help other people reach their full potential. I never spent one-on-one time with either of them, but I sat in convention halls and listened to them speak. They did more than inspire me to become a better leader; they helped me become a better person.

We have all known great team leaders in the athletic world. But we really only know the image their managers and the media want us to see. I knew an incredibly successful businessman who had the opportunity to have a superstar athlete endorse his product. After contemplating it, he chose not to go in that direction. When I asked him why, he said, "I've had a celebrity endorsement before, but he ended up getting in trouble and destroying his image, and it hurt our brand because my product had his image all over it." He went on to explain that just because someone can dunk a basketball, or hit a baseball, does not mean they are worthy of representing his brand. In his words, "It takes more than the gift of athleticism—it takes character."

Throughout the years, there have been many examples of incredibly gifted athletes who have worked hard and excelled in their sport but who have fallen short in other areas. I am not judging any of them; I'm just making an observation. Society often relates team-based leadership to a dominant athletic team, but that is probably because of the hype that revolves around televised sports. Our profession is quite different. As firefighters, in society's eyes, we are all lumped together, like soldiers. But for fire service leaders, there is much more to the picture. You have a team that you are responsible for. You are concerned with helping that team reach peak performance. You are serious about setting and achieving a higher standard for yourself and for those around you. You know changes need to be made, and you are up for the challenge. You know that a person's title does not mean as much as his or her desire to create positive change. If those last few sentences are true, congratulations for being exactly what our profession and today's society needs.

Leadership 101

A leader is a person who influences change. Leaders embrace the idea, stepping away from their comfort zone because they understand that leadership is about action, not position. Leaders do not rule like tyrants; they lead people to a better place. Chief Alan Brunacini once said, "Any idiot who outranks everyone in the system (like a fire chief) can line everyone up and yell, 'March!' But can that person really take them anyplace worth going?"

That's a good question to address. Are you taking your team anyplace worth going?

Leaders understand the value of personal development. They study their craft and resist the temptation of becoming complacent. They develop high-performance adaptive skills to be able to function in high-pressure scenarios, including skills like situational awareness, decision making, communicating, and negotiating. They understand their habits and actions create their reputation, and that the reputation of leadership is not given. It has to be earned.

There is so much to learn about leadership that sometimes it becomes confusing, but the act of leadership is actually very simple. In its purest form, leadership can be summed up in four steps:

1. Identify a problem.
2. Assemble a team of people.
3. Develop a solution together.
4. Solve the problem.

After that, reward your team and repeat the process by solving another problem. If you do this enough times, you will earn the trust of those you are leading.

> *Any idiot who outranks everyone in the system (like a fire chief) can line everyone up and yell, 'March!' But can that person really take them anyplace worth going?*
>
> —*Chief Alan Brunacini*

Are You Ready for a Leadership Role?

A firefighter who worked on an engine his entire career had recently been promoted to the rank of lieutenant. On his first day with his new responsibilities, another officer on the shift called in sick, and the lieutenant was assigned to the ladder company. A few hours into his shift, his company responded to a heavily involved warehouse fire. Their job was to set up an elevated stream and pour water on the fire from a distance. It was a defensive attack.

He told the driver where to position the apparatus. The driver had worked on a ladder company his entire career and had been on the job for the same number of years as the lieutenant. He was one of the most experienced aerial ladder operators on the job. Having raised and positioned the aerial hundreds of times, he felt the suggested position was incorrect, so he did what a good driver is supposed to do—he spoke up.

"Lieutenant, the stream will not reach from this location. We should move the apparatus in about 20 feet," the driver said.

The lieutenant heard the driver's opinion, but decided to discard it. "Position it where I tell you," he sternly ordered.

The driver and the other two firefighters on the ladder company did what they were told. After raising the aerial and establishing a water supply, they began to flow water, and missed the fire by 20 feet.

This newly appointed officer made a classic mistake of confusing position with experience (as discussed in number 7 below). Mistakes like this happen, but people in leadership positions need to learn from them; otherwise, they may not be ready for a leadership role.

Not every member of a team is ready to lead. In fact, some departments have a bad habit of allowing uneducated and sometimes incompetent individuals to be elevated to the highest level within their organization. This needs to be corrected. A fire service leader must be educated, competent, and prepared.

You may not even know for sure if you are ready to take on the role of a team leader. I believe you are, otherwise you would not be reading this book. Besides, I think one of the best ways to develop your leadership skills is to take action before you feel ready. We learn, grow, and change by doing, not by thinking about doing. To lead a team, you will also need other people who are ready to lead at various levels within your organization. To successfully lead, you will have to identify your key players—your critical few. Leaders have distinguishing characteristics, and so do people who are not ready to lead others. It is important for you to be able to determine who is ready to take on a leadership role and who is not. The following list can help you make that determination.

Ten Signs a Person May Not Be Ready to Lead

People who are not ready to lead may be characterized by any of the following.

1. They cannot handle pressure. Leaders are often admired, but they are also often criticized and sometimes even ridiculed. A true leader understands this and is able to put it in perspective. Sidney Greenberg once said, "A successful man is one who can lay a firm foundation with the bricks that others throw at him." In the fire service, it is not difficult

CHAPTER 3　　　　　　　　　　　　　　　　　　　　　　　　　　　　*Leading Teams*

to see how someone handles pressure. We deal with high-pressure situations on a regular basis. When people take on leadership roles, however, the pressure intensifies (fig. 3–2). They are no longer only responsible for their own actions. Now they are responsible for the actions of a team. Take note of how these people handle challenges and controversy. Do they play the victim? Do they blame others? By definition, victims are not leaders. Leaders step up and take charge during difficult times.

Fig. 3–2. Leaders must be able to make quick decisions while under pressure.

2. They have inadequate people skills. Most people who have inadequate or poor people skills do not know they have these shortcomings. Those who have strong people skills have the ability to make other people feel important. They are the type of people that others enjoy being around. They listen, pay attention, offer advice, and lead by example. There is a popular saying about leaders who are committed to improving their skills with people. That saying is, "Leaders are readers." One of the best ways for people to improve their relational skills is by reading the right books. When you see team members reading self-help and personal growth books, you know they are headed in the right direction.

3. Their values do not align with those of the organization. Leaders are culture creators. They are passionate about developing the right culture. Passion comes from values and beliefs. If a person clearly does not believe in the mission of the team, he or she is not the right person to help develop the culture you are trying to create. One way to determine if a person believes in your mission is to identify if the individual has been a good follower. Often good followers become good leaders because they have learned to adopt a vision and move in that direction for the betterment of the team.

4. They do not like being uncomfortable. Leadership is about change. The fact is, if you are not achieving your desired results, change is the only option. Change is uncomfortable. There is no way around it. When people avoid change, they are trapped in their comfort zone. No one will ever achieve anything significant without stepping outside of that familiar zone and into unfamiliar territory. Leaders realize the only constant is change.

5. They put personal needs ahead of the team. When people refuse to give credit where it is due, they are not ready to lead others. Leaders know that for them to become effective, they have to consider themselves the least important members of the team. When a person tries to take the spotlight all the time, he or she is showing signs of immaturity and selfishness. Others recognize this flaw immediately and tend to hold back from giving their best performance. A strong leader will understand that everyone on the team has needs and will try to accommodate those needs along the way. Leaders realize that the shortest, least significant words in the English language are *I* and *me*.

Drop the ego. No one works for you. They work with you. They work for the department.

6. They set bad examples. This one is obvious but needs to be mentioned. Even if people are strong producers, if they set bad examples along the way, others will follow those examples. Take a person who uses foul and offensive language regularly. When he reaches a leadership position, everyone on the team will think this language is acceptable. If team members become comfortable talking this way around the lunchroom, it is only a matter of time before they say the wrong thing at the wrong time outside of the workplace.

7. They confuse position with experience. Title, position, and rank mean something significant and should not be diminished. Even so, people in leadership positions need to stop acting like they have more experience in every area than the other members on their team. The fact is we can all learn something new from every member of our team. Even if you do have more experience than most, do not discount your team's personal experiences, which may provide valuable insight when you need it. The fire service does not need more officers who confuse position, title, and rank with experience; we need more leaders (fig. 3–3).

8. They are inconsistent. Team leaders cannot afford to be on one day and off the next. Yes, personal challenges will occur, and there will be times when people are not on top of their game, but that is different. When people are inconsistent as the norm, they are not stable enough to lead a team on a regular basis. This quality, and all of the others listed previously, can be overcome as long as the individuals do not exhibit the next two traits.

Step Up Your Teamwork

Fig. 3–3. The fire service needs more leaders.

9. They are indecisive. Indecisive people can be described as not having or showing the ability to make decisions quickly and effectively. Leaders make decisions. They do not sit on the fence and wait for everything to be perfect. They look at the options, make a decision, and act upon it. When a person takes too long to make decisions, rules often change and opportunities are lost. The "politically correct" thought process of "I feel strongly both ways" can take its toll on a team of people who are hungry for a clear decision and a precise action plan.

10. They are unable or unwilling. When people are unaware they are doing things incorrectly or are showing poor judgment, sometimes all they need to correct things is a little guidance. The real problems occur when someone is unable or unwilling to take corrective action. Inability can be overcome—most of the time—with training and effort. Unwillingness, however, is a completely different problem. People who are unwilling to do the things that need to be done have no place on your

CHAPTER 3 — *Leading Teams*

team and certainly should not be put in a leadership role. In the fire service, we have a word for people who are "unwilling": *insubordinate*.

Too many people in leadership positions are like travel agents. They are trying to send people to a destination they have never been to themselves. Don't be a travel agent. Be a tour guide.

Before you accept a role that will require personal responsibility and accountability, ask yourself, "Am I ready for a leadership position?" Do the same when considering a person for a leadership role on your team. If that person has a few of the characteristics listed above, address them or reconsider the appointment for the benefit of the team.

Remember, most of your skill set will come from doing. If you feel you exhibit some traits of a person unqualified for leadership, the key to growth will be to acknowledge the weakness. Look around and find people in leadership roles you admire and study them. Emulate the things they are doing that impress you, but do not lose your personal flair or style. Keep in mind, leadership is not always about being the one out in front. People often underestimate the importance of being a good follower first.

The First Follower

Leadership is born from being a follower. Do not expect perfection from your leader; you will end up being disappointed more often than not. The key is to learn from what a person does right, as well as from what he or she does wrong. When it is your turn to lead, do not be upset if the world is not ready to immediately follow you. Hopefully you have already earned the respect of your peers, but now is not the time to rest on your laurels.

Sometimes you have the advantage of being able to handpick your team members. For example, if you are hiring people to fill positions, you can look to hire those who play at what you work at. In other words, their strengths are your weaknesses. So much of a team's success comes from choosing the correct people, which is why I enjoy watching NFL draft day. You can almost predict if your team is going to have a good season based on its offseason picks.

Unfortunately, most of us do not get to choose who we work with. We have to make things work with what we have. Sometimes this presents a challenge for a young officer. Often I am asked by newly promoted officers how they can get the rest of their organization to buy in to their vision. I always respond the same way, "Don't try to convert the entire organization right now. Focus on one." You see, it takes at least two to make a team. And one of them has to have the courage to lead. That one is you. Now, you need someone who has the courage to follow. Share your vision, encourage others to help you create positive change, and begin to assemble your team. When you find one follower, value, respect, and treat that person as your equal. Make it about the mission, not about you.

The next step is to expand your team with another teammate, and then another, until your team is complete. If you do things right, the people who are not part of the team will feel left out and will want to become part of it. Embrace the people who want to become part of the mission. If you alienate them or take them for granted, you will fail as a leader.

It is a terrible thing to look over your shoulder when you are trying to lead, and find no one there.

—*Franklin D. Roosevelt*

Change Is Necessary

Change has become a very familiar word in the fire service. The majority of high-level officers in our industry agree that this is the most difficult time in history to be a fire chief. Most firefighters in America would say that one or more of the following statements relate to their organization:

- Tax revenues are down.
- We are not keeping up with technology.
- Our staffing level is the lowest it has ever been.
- Our operating budget is shrinking yearly.
- They are not replacing retired, educated firefighters.
- Our department is looking into the option of regionalization (shared services).
- The only way we can get the equipment we need is through grant funding.
- We are constantly being asked to do more with less.
- Our new management staff is leading us in the wrong direction.
- Newly appointed politicians are not supportive of our organization.
- Our SOPs are insufficient or outdated and need to be revised.

There are so many issues that need to be dealt with on a daily basis that the fire service has become a pressure cooker of stress. Still, the strongest leaders in our industry remain calm despite all that is going on around them. They focus on the solutions, not the problems. They learn to embrace and even create change. One of my favorite quotes is, "If you don't create change, change will create you."

I learned early in my career that most firefighters hate change. When an ambitious member of our team proposes an idea and asks why we are doing something that is not working, we need to stop responding with the terrible answer, "Because we've always done it that way" (fig. 3–4). We need to start considering different options. We need to embrace change, even if it bothers the dinosaurs. Robert Kennedy once said,

"Progress is a nice word. But change is its motivator. And change has its enemies." If you are leading a team, change can be frightening. You may be wondering if you are making the right decisions. Perhaps you are spending so much time weighing your options that you are not making any decisions at all. If either of those last sentences applies to you, the first question you will have to ask is, "Am I getting the right people involved in the process?"

Fig. 3–4. Why are we doing something that doesn't work? Is it because we've always done it that way?

Do not make the mistake of trying to do it all by yourself. You are surrounded by people who should share the same common organizational success goals that you do. Maybe you are not the head of the organization, but the head of a team within the organization. This does not mean you should sit back and wait for change to occur. Instead, show others that you and your team are willing to show initiative.

Whatever position you hold, your attitude toward change can tremendously influence the outcome. If your organization is like most, it is full of smart, talented people with flawed leadership skills. Many people have never had any form of leadership training, and even those who have had training will have to get past their own resistance to change.

What do you do when change, and especially "tough" change, is inevitable? Since this is happening more frequently, it is important to understand your role as a leader if you expect to survive and thrive. Change will require a leader to deal with various people in and out of the organization in different ways. Because of this, there are several different roles you must play to successfully lead thorough change. Leaders of change are self-educators, advocates, role models, decision makers, communicators, motivators, unifiers, commanders, and counselors.

If you don't create change, change will create you.

—*Unknown*

Change Can Be Good

I remember receiving a call one Christmas Eve from a friend and veteran firefighter who was distraught because he just found out he was being transferred from one shift to another. He had been working with the same group of firefighters for nearly 10 years, so when he received his notice, the fear of the unknown began to take over. To make matters worse, he had somewhat of a bad history with the man who was about to become his new superior officer. During our conversation, this firefighter actually said to me, "This is the worst thing that could possibly happen to me in my life." Considering the fact that he was married and had children, I thought that statement was a bit extreme. Nevertheless, it was clear that he was not open to this forced change.

I spent time talking to him about looking at this from a different point of view. I explained that this could be a great opportunity for him to break out of his comfort zone and grow as a firefighter. As I mentioned previously, I believe that a person, and a team, cannot grow inside their comfort zone. Growth only occurs when you step away from familiarity and into uncharted territory. After a long conversation, he finally agreed to go in with open mind.

About two months later, I ran into this firefighter at a social event. I asked him how it was going. He told me that he approached the situation from day one as if there was no negative history between himself and his new supervisor. He showed his new officer nothing but respect, and to his surprise, he was given the same in return. He went on to tell me that the two had genuinely become good friends, and I was surprised when he said, "Honestly, it's the best thing that could have ever happened to me." I reminded him of our conversation just a couple of months earlier, and we shared a laugh.

Most of us fear change because we do not know what it is going to be required from us. By changing, you may have to learn new skills, or meet and work with new people. Just those two possibilities alone can result in elevated levels of stress and anxiety, and justifiably so. We fear change because we fear the unknown, but we cannot let fear stop us from taking the necessary steps in order to move forward (fig. 3–5).

The fact is, if you are not achieving your desired results, change is the only option. It is an absolutely necessity, whether you like it or not. Change may not be easy, but if you do not like change, remember that you will surely like failure and irrelevance even less.

When change is in order, get to the heart of what needs to be changed. Do not just change what is easy; change what is necessary. It is one thing to recognize when change is needed, but as the leader of a team, it is your responsibility to inspire that change. Following are some tips to help you lead others through changing times.

> *The fact is, if you are not achieving your desired results, change is the only option.*

Fig. 3–5. Firefighters are used to stepping toward the unknown on the fireground. They need to be willing to do the same when change is necessary. *Courtesy:* John Harrison.

How to Lead an Organization Through Change

There are seven things to remember when leading an organization through change.

1. **Focus on the critical few.** I once heard a colleague jokingly say that change would be easy if it was not so hard. Although I do not fully agree with that sentiment, I do believe that leading a team through change would be easy if it was not for a select group of stubborn people who resist change. It may be true that most people do not like change, but it is also true that for every ten people on your team, there will generally be one who has the ability to influence the others. One of the keys to implementing change is to first recruit your top

influential team members and utilize their ability to influence and inspire others.

2. **Paint a crystal clear picture.** You want to be clear about what it is that needs to be changed. Bring your team together and be as clear and concise as possible about what it is you intend to change and why it is necessary to do so. Talk about the benefits, not just for the overall team, but for the individuals who help you accomplish your goal. Remember, people have needs, and they will be more excited to know that their efforts will help them accomplish their personal goals as well. Make it clear to all that new worlds emerge when old patterns are broken.

New worlds emerge when old patterns are broken.

3. **Bury the game ball.** Forget about past failures. Your rearview mirror is smaller than your windshield for a reason. Where you are going is more important than where you have already been. Do not dwell on what has not worked. If you do, you will end up wasting valuable time and energy that would be better utilized working toward the solution. Move on with the intention of winning, and continually remind yourself that your team cannot stay in one place. They are either moving forward or backward.

4. **Keep it simple.** Have a simple plan and lead with a simple message. Charles Mingus said, "Making the simple complicated is commonplace; making the complicated simple, awesomely simple, that's creativity." Don't make the common mistake of overcomplicating a simple task. According to CollinsDictionary.com, the term simplexity is defined as "the process of simplifying something by obscuring the more complex aspects of the original goal." In other words, it is a process in which a simple correction or improvement results in a more complicated system. This concept, which has been adapted in advertising, marketing, and other industries, has also found its way

into the fire service lexicon. Do not make the mistake of thinking the better way always has to be the more complex way.

5. **Encourage others.** To encourage is to empower. When a salesperson makes 50 calls without closing a sale, that person needs to know that number 51 may be the one that makes all the difference. The same goes for an athlete who missed a game-winning shot or the firefighter who failed to contain the fire. No one has a 100% success rate. Even the best major league baseball hitters in the world fail to make it on base seven out of every ten times at bat. Prepare your version of a game day speech and remember that true stories of triumph over adversity inspire people to believe the impossible is possible. It does not take more than a few minutes on the Internet to find stories of teams that have overcome adversity and achieved the type of success that you are trying to achieve. Share them with your team as examples of the good that can come from change.

6. **Take action.** You've heard the saying "Strike while the iron is hot." If you subscribe to that way of thinking, perhaps you should stop waiting for the iron to become hot, and make it hot by striking. When you take action, great things can happen. And if you are hesitant to jump in with both feet because of fear of the unknown, take Frederick Wilcox's advice. He said, "Progress involves risk. You can't steal second base and keep your foot on first."

7. **Measure your progress.** Without a mechanism for evaluation, you cannot measure your progress. When you meet with your key players, determine how you will evaluate the new methods your team initiates. After initiating your plan, compare the past results with the present results. Review every aspect, evaluate your options, and revise your methods if necessary.

You can't steal second base and keep your foot on first.

—Frederick Wilcox

Keep in mind, if you are not achieving the results you desire, you are either doing the wrong things or not enough of the right things. Once you determine which it is, you can work on changing your tactics and moving forward. For things to change, you have to change. Do not fear change, because when you really take a moment to think about it, change is the only constant.

Transparency

Asking a team to change without being transparent about the direction you are heading does not work. As an author, I have written quite a bit about leadership traits. I even dedicated an entire chapter of *Step Up and Lead* to that very topic. I listed the 13 top traits associated with leaders in the fire service. These traits can be remembered by using the acronym LEADERS TEACH, which stands for the following:

1. **L**oyal
2. **E**ducated
3. **A**daptable
4. **D**etermined
5. **E**nthusiastic
6. **R**eliable
7. **S**elfless
8. **T**ough
9. **E**mpathetic
10. **A**ssertive
11. **C**ourageous
12. **H**onorable

The 13th trait is the acronym itself: LEADERS TEACH what they know to others.

I once spoke to several hundred firefighters at the Tri-State Fireground Operations Seminar at Kean University in New Jersey. I was honored to share the stage with a fire service icon, Deputy Chief

Vincent Dunn (FDNY). The topic of my presentation was "Leadership in the Fire Service." Chief Dunn knows a little something about leadership (fig. 3–6).

Fig. 3–6. It was an honor to be able to meet, share the stage with, and spend time with Deputy Chief Vincent Dunn (FDNY) (right), a true fire service leader.

During the seminar I stressed the fact that there is a need for strong leadership in today's fire service. I spoke in detail about each of the traits listed above. After the class, a group of dedicated firefighters approached me to speak about one of the biggest problems we are facing with leadership: the lack of transparency. These firefighters were complaining about their chief in a very open and candid manner. They were saying things like, "He's always having secret meetings," and "He never takes our advice. He acts like he's listening, but he always looks up at the clock or takes phone calls when we are discussing important issues." They even said, "We have had meetings with him where we all collectively agreed on a direction to take. Two days later, he sends down a policy reflecting the complete opposite direction we agreed upon." These types of incidents happened frequently. One of them concluded by saying, "He has a secret agenda and none of us know what it is. That scares the heck out of me."

During our conversation it became apparent to me that transparency should be on that list of essential traits of leaders. This holds true outside the fire service as well. Transparency is essential if you want to lead any team.

It seems that our country lacks leaders with integrity. Just ask 10 people if they plan on watching the next political debate, and at some point in the conversation, 9 out of 10 (if not 10 out of 10) will say something like, "What's the point? They all lie anyway." Consider past presidents (and presidential candidates) who have said, "I will never raise taxes." We want to believe them, but we all realize it is likely just lip service.

People sometimes ask me if I am a Republican or a Democrat, and I used to answer, "I'm neither. I vote for character." Well, I cannot remember a politician at the national level that I believed was in politics solely for the right reasons. I have learned through the years that the only thing you can trust about a politician is his or her past record.

Transparent leadership is needed now more than ever. If you are currently in a leadership position, whether as a parent, coach, CEO, or fire officer, you owe it to your team to be a leader with a clearly defined goal and mission. People do not like workplace surprises. They value transparency in leadership because it allows them to work free of distractions and to make better choices. Leaders who are transparent and take time to explain what they are striving to accomplish will find it easier to get others to participate in their mission and believe in them. If you want your team's full involvement with what you are trying to accomplish, they have to trust you. Trust comes when you share information openly.

Transparent leaders are real. They recognize their strengths and weaknesses. They understand that their actions impact the actions of others. They are willing to admit mistakes so they can address problems. If your goal is to have a team of employees who genuinely want to follow you, as a leader you must acknowledge the reality of each situation, good and bad, so that together you and your team effectively respond to the bigger issues you will encounter. Team members become detached when they are ignored, not taken seriously, stereotyped, or left in the dark.

Transparent leaders are confident with their leadership decisions and strategy and are true to themselves. I would even go as far to say that transparent leaders live more fulfilling lives because there is no secret agenda.

Do not confuse transparency with radical transparency. *Radical transparency* is a management method where nearly all decision making is carried out publicly. All draft documents, all arguments for and against a proposal, all final decisions, and the decision making process itself are made public and remain publicly archived. This is not always a good thing. There are times when a leader, or team of leaders, will have to iron out some kinks before presenting or proposing a potential plan of action to the troops.

One reason why some leaders are not more transparent is that they believe they will be viewed as less authoritative. Another reason could be because they may be in over their heads. Often when people in a leadership role do not know what actions to take, they take no action at all. When this happens, people automatically think something is up because no one is saying or doing anything. When a fire department is low on staffing and the members are worried about layoffs, demotions, or dismantlement, they need a team leader who is transparent. Someone who is standing in front of town officials saying, "Enough is enough!"

A friend who was a former training officer for another department once shared this story with me.

> Last week, I spoke to my chief about promotions. We have an opening for a captain, and I didn't understand why he isn't pushing to fill the spot. I asked, and he said the town council will not allow any promotions at this point. We have a good relationship, and I know the candidate sitting at #1 would make a great officer. I urged him to fight for the position so we didn't lose it. He promised me that he was fighting as hard as he could.
>
> A few days later, I was sitting at the desk outside my chief's office organizing drill sheets when his phone rang. He picked it up and acknowledged the council member on

the other end. I could hear the call as clear as day because his door was open, and he left the call on speaker phone. The council member actually came right out and asked if the chief was going to request putting a promotion on this month's agenda. My chief replied, "I don't need another officer right now. I have nowhere to put him."

This was a blatant lie. He absolutely does have an opening. The ONLY thing I can think is that he either doesn't like the guy on the list, or he had a secret agenda that he doesn't want to tell anyone about. Now I'm sitting here wondering what else he's been lying to us about.

Would you want this type of person to be the head of your organization or team? Lack of transparency can be translated as an absence of trust. Trust is one of the most valued qualities among great team leaders.

Colin Powell, former American diplomat and retired four-star general in the US Army, was the 65th US Secretary of State, serving under President George W. Bush from 2001 to 2005, and the first African American to serve in that position. He also authored a book titled *It Worked for Me: In Life and Leadership*.

Powell was once asked the question, "How do you define the key characteristics of effective leadership?"

His immediate response was, "Trust. The longer I have been in public service, the more people ask me about leadership. Leadership ultimately comes down to creating conditions of trust within an organization. Good leaders are people who are trusted by followers. Leaders take organizations past the level that the science of management says is possible." Powell went on to say, "They'll follow you into the darkest night, down the deepest valley, up the highest hill if they trust you."[1]

Trust is something you must work for and earn from the start. It is not something you want to try and earn after you develop a bad reputation. When you are the new kid on the block, earning trust is your first order of business. If you are new to a team, earn trust now, and when you are put into a leadership position, you will not have to try to recreate yourself.

Transparency, trust, and honest communication allow for quicker alignment to the goals and vision you have for the team you are leading. If problems are identified and solved faster, your team will not feel like they are constantly starting and stopping.

There are five things that can happen when a team leader is transparent.

Five benefits of transparent leadership

1. Problems are solved faster. Everyone learns more about each other and can focus solely on solving problems when their leader is transparent. For example, if the head of an organization is constantly having secret meetings and planning big changes without the team's knowledge, there may be programs and initiatives the team has been working on that will never be possible because they will not align with the new (secret) mission of the organization. If the leader is open about the direction the team will be heading, they can all work together because now they know what problems they have to solve.

2. Teams become stronger. Transparency is a powerful unifier. With a transparent foundation, the leader can openly discuss the strengths and weaknesses with the team. When weaknesses are identified, the team can work on strengthening them. They can also strategically match people with specific tasks based on their individual talents, skills, and ability. Transparency allows relationships to mature faster, as openness can potentially avoid misunderstandings that can fuel unnecessary tension.

3. Relationships grow authentically. Transparency brings people together. When like-minded people begin to learn more about one another, they will naturally be drawn together. Each team member will also learn to value the others for what they bring to the table. When job-related problems arise, you will know which person to assign to tackle specific tasks. As the team learns more about each other and begins to see how they are alike and different, they can begin to recognize how they complement one another.

Step Up Your Teamwork

4. Team members begin to trust their leader. When leaders are transparent, especially during challenging times, they will earn the trust and respect of their team, who will become great supporters in time. It is important that when new members come on board, the existing team members are able to promote their leader as trustworthy, competent, and honest. This eliminates any negative preconceived judgments that others may have had about the leader. If you are in a leadership position, you will accomplish more when people come on your team and immediately trust you because of your reputation.

5. Teams perform at a higher level. When doubt is eliminated, your team can focus on performance and service to others. Never forget why your team exists. If they spend all their time wondering about the secret motives of others on the team—especially the leader—they cannot become a high-performance team. It really is that simple.

Be open and honest with your team and encourage others to be the same. When developing your leadership traits and style, put transparency at the top of the list.

> *How do you get the respect of the men?*
> *Be honest, be fair, and be consistent.*
>
> —Major Dick Winters, 101st Airborne
> (Band of Brothers)

Transparency in the Digital Age

You do not have to be sitting outside of a person's office when a call comes in to learn about the individual's secret agenda. Social media has suddenly given people permission to enter a leader's personal space, a place they were previously prohibited from entering. The digital age has changed the levels of transparency that we expect from people. There is a reason people would rather watch a video blog than read a written blog. Being able to view someone's facial expressions and body language goes a long way. Subconscious gestures and movements help us evaluate the authenticity of others. But even if someone does not record personal videos, you can still get to know what they are really all about just by reading their daily status updates or tweets. Instead of a business card, when I meet someone in a leadership position, I ask if they have a social media page that I can connect with them on. I want to see where their head and heart is at. I want to see how transparent they are. If they score high enough on the transparency scale, I know they have a key ingredient in what it will take to develop a loyal team.

Specific Intent

A transparent leader has another advantage over those who have a hidden agenda. If you want to develop true synergy, you will want a team that is conditioned to be disciplined and specific. This is not possible without a clearly defined mission. *Specific intent* is a simple concept to understand and measure. All you have to do is ask yourself, "Is what we are doing or about to do getting us closer to our goal and objective?"

Imagine 15 firefighters arriving on scene of a working structure fire without any specific plan. They will fail at their mission of saving property and life. Your team cannot afford to throw mud at the wall just to see how much of it will stick. You have to have a plan of action, and everyone needs to know their job. Every team member needs to take action with specific intent, starting with the leader.

Step Up Your Teamwork

Is what you are doing or about to do getting you closer to your goals and objectives?

Understanding People

It will be impossible for you, as a leader, to develop a high-performance team if you do not learn the *how* and *why* of motivating people (fig. 3–7).

Fig. 3–7. Leaders must learn *why* and *how* to motivate their team members. *Courtesy:* Cindy Rashkin.

Your ancestors may remember going to the county fair back when one of the biggest attractions was trained fleas. If you do not know what I am talking about, or you do know but refuse to admit you were around back then, let me explain what they are. Imagine you are 10 years old, walking through the fair, probably on your way to get some cotton candy or ride the Ferris wheel, when you notice a large crowd gathered around one of the booths. You approach, only to discover that everyone is paying a dollar and jockeying for position to get a chance to peek into a jar full of fleas. Although the jar does not have a lid on it, the trained fleas do not hop out; they just spring around, never leaving the confines of their imaginary prison.

How was this possible? The answer is quite profound. The flea trainer places the fleas in the jar and puts the lid on it. As the fleas try to jump out of the jar, they continuously smash their little heads against the lid. After a while, they realize they cannot get out that way and ultimately stop trying. They are trained to believe that they can only go so high, and no further. At that point, the flea trainer removes the lid and collects your dollar.

The problem with many ineffective teams is that somewhere along the line, the members of those teams were also told, "So high, and no further." We live in a society that is sometimes so self-absorbed that people forget to pay attention to those who are around them on a daily basis. Some of you reading this book may have even neglected your children or spouse at some point, and it was not because you did not care about them. It was because you were so caught up in the minutiae of daily life that you lost sight of what was important. We have all been there. If we take time to learn about and understand what motivates people around us, we can avoid many of the problems that have kept us, and our teams, stagnant.

There was a firefighter named Manny (not his real name) on my team who was very disinterested in the job. I would even say he was often bitter and angry when he would come to work. Manny had been at odds with his captain, and the tension between the two was obvious. One day, I brought the captain into my office and asked what the deal was. He told me that Manny was difficult to work with and often made disparaging comments about the captain under his breath to the other firefighters. After talking with the captain, I met with the senior man on the group, who spoke candidly with me about Manny, "Manny thinks he should be an officer and he's disgruntled because he thinks he has more to offer."

The next meeting I had was with Manny. I brought him into the office with the captain, who knew what I was about to do because we discussed it earlier. We spoke for a moment, and then I said, "Manny, I need your help." He perked up.

I continued, "You are a talented firefighter with a tremendous amount of knowledge about the job, and we have three probationary firefighters who need the best training we can give them. I know you're a team player, so I'm sure I don't even have to ask, but I will anyway. Would you work with the captain to develop a drill that you and he can give on elevated stream operations?"

Manny's chest expanded. "Absolutely. I'd love to!" he replied.

The rest of the day, Manny had extra pep in his step and was obviously feeling good about himself. The best part was that the relationship between the Manny and his captain significantly improved from that moment on. The experience made me realize something very important. Most people will run a little harder for the rest of the day after one good compliment. I also realized that encouragement is as vital to the soul as nutrition is to the body.

Most people will run a little harder for the rest of the day after one good compliment.

The flashpoint

Have you ever seen someone's passion ignite right in front of your eyes? The person's eyes widen like a kid who has just seen Mickey Mouse for the first time. It is the moment of discovery that there is a very real possibility that a particular achievement or task that the person is passionate about can be accomplished. A dream is becoming attainable. Something awesome happens inside a person at this moment. It is like a switch is turned on and a new energy is released after years of restriction. I call this a person's *flashpoint*.

In firefighting terms, a *flashpoint* is the lowest temperature at which the vapor of a combustible liquid can be made to ignite momentarily in air. In layman's terms, it is the point at which eruption into significant action occurs.

Symbolically, this phenomenon can also happen within people. A person can reach flashpoint when the following three things occur:

1. The person's dream or mission in life is clearly defined.
2. The vehicle that is going to help make that dream a reality is found.
3. The person takes action and begins seeing results.

Team flashover

There is an even more amazing phenomenon that occurs when several team members reach their flashpoint simultaneously. I call it *team flashover*. In firefighting terms, a flashover is defined as "full room involvement." This is where synergy and momentum kick in. Team flashover, or full team involvement, should be your goal if you want to achieve true success as a team. Let's talk about how to bring all the pieces together.

Find their hot button

To inspire your team, you will need to discover what motivates their behavior. According to humanist psychologist Abraham Maslow, our actions are motivated by a hierarchy of needs (fig. 3–8). In his 1943 paper, "A Theory of Human Motivation," published in *Psychological Review*, and in his subsequent book, *Motivation and Personality*, Maslow suggested that people are motivated to fulfill basic needs before moving on to other, more advanced needs.

Maslow's hierarchy of needs is often portrayed in the shape of a pyramid, with the largest, most fundamental level of needs at the bottom. He believed these needs are similar to instincts and play a major role in motivating behavior.

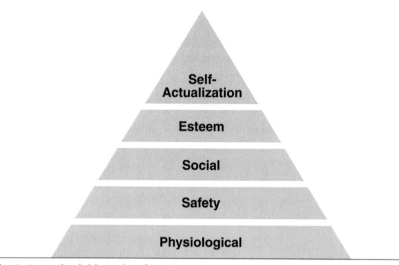

Fig. 3–8. Maslow's hierarchy of needs

There are five different levels in Maslow's hierarchy of needs:

1. **Physiological needs.** These include the most basic needs that are vital to survival, such as the need for water, air, food, shelter, warmth, and sleep. Maslow believed that these needs are the most basic and instinctive needs in the hierarchy because all needs become secondary until these physiological needs are met.

2. **Security needs.** These include needs for safety and stability. Security needs are important for survival, but they are not as demanding as the physiological needs. Examples of security needs include a desire for steady employment, health care, safe neighborhoods, and greater shelter from the environment.

3. **Social needs.** These include needs for belonging, love, and affection. Maslow described these needs as less basic than physiological and security needs. Relationships such as friendships, romantic attachments, and families help fulfill this need for companionship and acceptance, along with involvement in social, community, or religious groups.

4. **Esteem needs.** After the first three needs have been satisfied, esteem needs becomes increasingly important. These include the need for things that reflect on self-esteem, personal worth, social recognition, achievement, respect, and accomplishment.

5. **Self-actualization needs.** This is the highest level of Maslow's hierarchy of needs. Self-actualizing people are self-aware and are concerned with personal growth. They are problem-solvers who are not as concerned with the opinions of others; they are interested in fulfilling their potential.

Although your goal is to have every member on a team commit to the overall team mission, do not lose sight of the fact that your team is comprised of individuals with personal needs.

What makes your team members tick? Every member of your organization has a *hot button*—something that elicits a strong emotional response or reaction. There are many common motivators that cause people to work hard and take corrective actions. For example, team members may be motivated by a desire to achieve any of the following:

1. Be healthy
2. Get out of debt
3. Pay for children's college
4. Enjoy better relationships
5. Have more time to pursue hobbies
6. Free a spouse from having to work
7. Vacation more
8. Take early retirement
9. Spend more time with family and loved ones
10. Donate time or money to charity
11. Make a difference in society

Learning what each of your team members wants is the single most important tip on how you can best motivate others. There will be times when their motivation and energy levels are low. The most effective way to provide them with a motivational boost is to remind them what they are striving for. There is a very specific technique you must use in order to discover what a person's dream, goal, desire, or hot button is—ask them!

Communicate with them in an open and sincere way. Let them know you are as genuinely interested in them and their needs as they are themselves. The moment they think you care more about your mission than you do about their well-being is the moment you start to lose their trust. Legendary UCLA basketball coach John Wooden was never concerned about finding basketball players who knew *how* to play the game. He was more interested in finding those who knew *why* they were playing the game. He understood that the person who knows *how* to do something will always be at the mercy of the person who knows *why* they are doing something.

I once had the good fortune of meeting the head of a fire academy who told me he always asked new recruits the same opening question: "Why are you here?" He explained to the recruits that "I like to help people" was not an acceptable answer for him. He already knew that, but the recruit could have chosen another service-based profession. He wanted to know why that individual specifically chose to be a firefighter. One individual told him it was because he had failed at everything else he had ever done. Unfortunately, the recruit also flunked out of the academy. A couple of days later, the failed recruit tried to commit suicide. When the academy head visited the man in the hospital, he asked the recruit why he did it, to which the recruit responded, "I told you I failed at everything. I needed to succeed at something." Although the individual did not make it as a firefighter, the academy head helped the man find a good job and achieve some much-needed victories in his life.

Another recruit, who was the son and brother of a firefighter, when asked why he wanted to be a firefighter answered, "I'm hoping my father will love me as much as he does my brother someday." This recruit's father spent a lot of time with his brother, and he craved that same level of love and affection. Choosing the profession of firefighting was a way of getting his father's attention and hopefully earning his respect and love. Clearly, people are driven by the *why* behind their choices.

Helping a teammate define his motivation is worth the time and effort it takes. It provides an advantage over teams who do not know each other's motivations (fig. 3–9). A volunteer fire chief may read this and wonder how he would be able to develop a stronger team by knowing what his members' personal goals are. The reason is simple. You are showing that you are a leader who genuinely cares about the things that are important to your team members. When I discover firefighters who are determined to advance their careers and become officers, I also ask about what is motivating them. Then on days when they seem disconnected, I can remind them what matters to them and help them get back on track.

Fig. 3–9. Team members who understand the *why* behind their choices have the advantage over those who do not. *Courtesy:* Brett Dzadik.

Once you know what motivates your team members, all you will need to focus on is what technique each of them responds to, and you're on your way to stepping up. Some will respond to direct orders, as if they were soldiers in the military. Others may be more sensitive and will respond better to a caring push. (I have known many firefighters who have responded better to this type of motivation.) I cannot stress enough how important it is to take time to learn how your teammates respond. It is no secret that people are different, but many managers, executives, and team leaders fail to recognize the fact that they cannot manage or motivate two different people the same way and expect the same results. Motivating people is an art form. While Billy Martin and Tommy LaSorda were both successful managers in Major League Baseball, there are many baseball players who could not have played for Billy Martin. His brash and direct methods simply would not work for many people. Tommy LaSorda, on the other hand, cared for his players in a way that many managers could not.

Remember, not everyone wants to be an officer. Take the time to find out what your team members want to accomplish. Maybe one wants to be a driver or would like to be assigned to the rescue company. Regardless of their goals, as a team leader, you must be supportive and encouraging. Team members will work harder knowing you approve and care.

I have never seen a man who could do real work except under the stimulus of encouragement and enthusiasm and the approval of the people for whom he is working.

—Charles Schwab

The Lost Art of Listening

Of all the skills needed for leading teams, listening is the least understood. The majority of people do not listen with the intent of understanding. They are not interested, or they do not seem to be interested in what others are thinking or feeling. Great leaders don't make this mistake. They never stop listening. They know that listening is the best way to get word of impending problems before they become obvious to others. They also know that opportunities come from listening, not from talking. Wise people will intentionally remind themselves every morning that they will not learn anything new by talking. If they are going to learn anything new at all, they will do so by asking questions to the right people and listening to the answers. To put it another way, God gave you two ears and one mouth, so use them proportionately.

One of the hardest things to do in life is to listen without intent to rebut. Too many people think that the key to leading a team is by always being right. This is incorrect and could prove fatal to a team. A great leader will want to hear what others are thinking. When people feel they are being listened to, they feel respected and valued. Henry David Thoreau said, "The greatest compliment that was ever paid me was when one asked me what I thought."

Once you recognize the fact that listening is one skill that is often overlooked, take an honest look at yourself and your leadership style. Ask yourself, "What do I really know about my team members?" There is a good chance that you do not know what their goals and dreams are, although that will change now that you have read this chapter. Do you know the names of each team member's spouse and children? Do you know where they live? You will learn these answers when you begin to ask questions that show you care about them. When you ask a question like, "How's everything at home?" they will appreciate that you realize they are individuals with their own lives outside of the fire station.

Six rungs of active listening

The more you listen, the more people will like you, and the better conversationalist you will become. To be an effective listener, follow the six rungs of active listening:

1. Look directly at the person who is talking.
2. Listen attentively and respectfully.
3. Do not interrupt the person who is talking.
4. Replay in your mind what you heard the person say.
5. Pause before you respond and ask subject-related questions.
6. Use the words *you* and *your*.

There is one piece of advice I would give any person who is sincerely interested in becoming a better listener: to read the book *How to Win Friends and Influence People* by Dale Carnegie. When I first heard that title, I thought it was about how to manipulate people to think the way you wanted them to. A friend told me I had the wrong impression and encouraged me to read and see for myself. I took his advice, and I can honestly tell you that book helped me to improve every relationship in my life, on both personal and professional levels.

Bonding

A friend who was recently promoted to chief officer called me up one day to talk about a challenge he was having.

"I work with a great bunch of firefighters, but they don't seem to get along with each other. They don't fight, but there's no team spirit, no bond, no unity," he explained.

He went on to tell me they drilled together daily, but when they returned to quarters, they each went off and did their own thing until an alarm came in. I told him that was not uncommon, but I became concerned when he said, "No one seems to enjoy being here."

"When was the last time you all got together outside of the firehouse and did something fun?" I asked.

"We haven't since I've been here," he replied.

I encouraged him to schedule some type of outing where they could all get together, meet each other's families, and cut loose. The firefighters on my department have come together regularly for many of these types of outings: fishing trips, softball games, picnics, pool parties, or pig roasts. He thought that was a great idea and called me a month later to share the results. He said he planned a rafting trip down the Delaware River, and three-fourths of his team showed up with their families. They had a great time on the river. Afterwards they went to a nearby lake, fired up the grills, and played some volleyball. I could hear his excitement as he described how great it was to see his wife and kids interacting with the other families. He was especially happy that people were smiling again around the firehouse. Their energy levels were up, and they had already scheduled their next get-together—a dinner in the Little Italy section of New York City.

No matter what your title or position on your team, make the effort to bond with others. Get to know the people you work with on a personal level so you see them as people and not as coworkers (fig. 3–10). Here are a few tips on how you can accomplish this:

1. Get your team together in an appropriate setting that is favorable to bonding.

2. Create opportunities for talking and getting to know each other.

3. Move forward with a respectful relationship, regardless of your differences.

Although I have always found the best bonding experiences to be those that occur outside the work environment, there are plenty of opportunities to develop stronger relationships during the course of an average workday. As a firefighter, I found it easy to bond with the guys I worked with day in and day out. This was in part because we had a lot of downtime between our training and fire calls. When I became deputy chief and my workload increased significantly, this became more difficult. However, I still made it a daily goal to visit each of our four fire stations and spend a few minutes talking with and encouraging my team. I enjoy asking how their families are doing or talking about where they went on vacation. Then we talk about what they have planned for the workday. I could not imagine ordering a firefighter to advance a hoseline into a basement charged with smoke without knowing him or her on a more personal level. My team needs to know I care about them and have their best interests in mind. Your team needs a similar reassurance from you.

Fig. 3–10. Firefighters from East Haven, Connecticut participating in a wellness challenge

Team bonding is a key ingredient to overall team building success. While working on projects and assignments, or training at work, people will begin to establish a rapport with each other. However, team outings allow them to take this further and establish friendships. As the leader, it is up to you to find ways to develop stronger relationships within your team. Do not wait for opportunities to fall onto your lap, but instead, take the initiative to create them, and your team will follow your lead. Possibilities could include a social gathering like a family fun day or a holiday party—just remember not to exclude anyone. Such events provide great opportunities for you to discuss something other than business and will help to strengthen the bond you hope to create among teammates. Brian Tracey says, "Eighty percent of life's satisfaction comes from meaningful relationships." Team bonding is a great way to develop those relationships. A team that plays together stays together.

At these team-building events, it helps if people avoid detailed conversations about politics or religion. As much as we like to think we are above such behavior, relationships have been severed over opposing political views, and wars have been fought over religious preferences. It would be a shame to have such a positive experience be short-lived because two individuals wanted to stress their personal points of view in a combative or confrontational manner. I speak from experience when I urge you to avoid this whenever possible.

Eighty percent of life's satisfaction comes from meaningful relationships.

—Brian Tracey

Give More Responsibility

When my son Thomas was nine years old, there were times when he would talk back to his mother or neglect to do his chores—behavior typical for boys that age. For a period of time, whenever I left the house, I warned him that he had better listen to his mother. Sometimes he listened, but not consistently. One day, I decided to try another tactic. After putting my uniform on, I asked him to walk me to the door.

At the door I took a knee in front of him and said, "Thomas, you understand that when I go to work, my job is to take care of people, right?"

"Yes," Thomas agreed.

I continued, "When I'm gone, it's really important that you take care of Mommy. I'm going to leave you in charge."

His eyes lit up. "Really?"

"Yes. Can I tell you what being in charge means?" I asked.

"Yes, Dad."

"Being in charge means you do the things Mommy and I have taught you. It means doing the right thing and making good decisions. Can you do that?" I asked.

"I can do that, Dad," he replied.

I do not know that I had ever seen such a proud look on his face up to that point in his life. It reminded me of a saying I always share at my leadership seminars—people will rise or fall to meet your level of expectations for them.

When you give responsibility to others, you are telling them that you have confidence in their judgment and ability. You are also giving them a chance to learn. Even if they fail at the task initially, they will gain valuable experience. You may not agree with the way someone approaches a task, but if he or she achieves the desired results that benefit the team, it is worth the exchange.

Get good at leading by example, but delegate tasks to others so they can develop their skills. Otherwise, you will end up holding your team members back. It has been my experience that people who feel contained (like those circus fleas in a jar) eventually leave that team for a better environment. When someone else hires them and gives them the freedom to shine, their previous employers may wonder what caused the sudden improvement in performance. The answer may be that they were not allowed to shine in their previous roles.

Establishing Team Expectations

When the team has high expectations and performs well, they feel like they can take on the world. When they provide exceptional customer service, coordinate with each other on projects, and work to help develop new recruits, you know you have a winning formula. But what if your team consists of lazy, unmotivated individuals whose personnel files have more disciplinary notices in them than continuing education certificates? What if they are rude to customers and disrespectful toward authority? What do you do then?

One of the very first things you should do with a new team is sit them down and tell them what you expect from them (fig. 3–11). One of the biggest problems managers and supervisors encounter in the fire service is self-imposed—they do not establish expectations for their team.

Fig. 3–11. It is imperative for leaders to establish team expectations and communicate those expectations early on. *Courtesy:* P.J. Norwood.

Why does your team exist? What is your goal at every incident? How can you make sure you go above and beyond to exceed your customer's needs? These are examples of things you want to talk about with your team early on. If they know where you stand and what you want, you will have a better chance of getting them to perform in a way that you find satisfactory. When I was assigned as the tour commander of my own shift, I spent the week leading up to my first day writing out what I wanted to discuss with my newly assigned team. My rough notes looked like this:

What I Expect from Tour C:

1. Train for no less than three hours a day.

2. Exhibit professionalism on the training ground and fireground.

3. Follow all the rules, regulations, policies, and acceptable procedures of the KFD.
4. Exceed our customer's expectations, *every time*.
5. Know your job and your equipment.
6. Know and preplan the streets and buildings in our town.
7. Recognize that self-education is as important as required drills.
8. Take care of the 3Fs: firefighters, fire apparatus/equipment, and fire stations.
9. Make proper risk assessments and safe operations a priority,
10. Talk to me, or your company officer, if you have any problems or concerns.
11. Make good decisions.
12. Be a problem solver, not a problem finder.
13. Take care of each other because we are a family.
14. First lead a team of one—yourself. Because if you cannot lead yourself, you cannot lead others.

The day I met with the team, I went down the list and elaborated on my expectations for each category. For example, on every shift, I required no less than 3 hours of training. I even broke it down to 2 hours of hands-on training, 30 minutes of preplanning buildings, and 30 minutes of EMS training. This was important; now they knew the absolute minimum of what I wanted. I set the team expectations up front. Now there was no guesswork. Doing more than what I required would be great; less would be unacceptable.

After giving examples in each area, I also told them what to expect from me and asked them to hold me accountable for my actions. Because we were now a team, it was important to me to lead by example. And I was not above them. We were a family, and I valued each and every one of them.

So now it is your turn to lead. You have a team and you are ready to lead, but you are not sure where to begin. Here are some things to keep in mind while establishing your team expectations. Whatever expectations you set, make sure they meet all of the following criteria.

Criteria for establishing team expectations

1. Legal, moral, and ethical. This may seem obvious, but you would be wise to remember the times we are living in are quite different from that of our parents. Laws change, and what was once considered morally and ethically acceptable 10 years ago may not be today.

2. Realistic and achievable. People want to be led, and they like to achieve more than they thought possible, but nothing can be more frustrating for a new team than having a newly assigned boss set unrealistic and unachievable standards.

3. Conforms to acceptable standards. In the fire service, we use standards from the National Fire Protection Agency as our guide. We also develop standard operating procedures. These standards exist for a reason. You can use these documents to help communicate your expectations. If you disagree with your current organizational policies and would like to change them, take the appropriate actions. Do not disregard them and do whatever you want. If you do that, others will do the same with your rules and expectations.

4. Consistent with the culture you are trying to create. Culture is created either by design or by default. If you do not tell your team members how you expect them to treat customers and how you expect them to operate at any given scenario, they will do what they want. When that happens, you are giving permission for each individual to create his or her own culture. This ends in team failure more often than not.

5. Fair in treatment of all team members. It is obvious when a team leader plays favorites. Matching individuals with tasks that their skills are suited for is one thing. Giving the guy or gal you went to high school with all the best assignments is another thing altogether. Be as fair in reward and recognition as you are in discipline.

6. Inclusive for all members of the team, regardless of rank. Do not just set expectations for your frontline people; set them for everyone. Tell the newest members of your organization that you expect them to step up because they have a lot to learn. Let the senior team members know you expect them to set the right example for the others.

Setting your expectations up front is a great practice to get into. Once you start doing this, you will understand how valuable this simple action can be. It may take time for you to work out exactly what you want, but it is time well spent. I can count on one hand how many times I have had to ask a member of my team why he or she was doing something completely different from what I expected. This is partly because I have a highly competent team, and partly because our team expectations were clearly established.

Priorities

What is your number one priority? My top priority is doing everything in my power to ensure my team members go home healthy at the end of each shift. If I do my job right, the men and women who work under my command will enjoy long, healthy, and hopefully happy retirement years. As a team leader, there are many ways you can inspire your team to better prepare for success. You may have to sit down and talk to them in depth the first time you set expectations, but then you will find ways to remind them of those expectations with simple phrases or words. For me, those words are, "Never walk past a problem you can solve. Take charge, and always do the right thing."

Never walk past a problem you can solve.
Take charge, and always do the right thing.

CHAPTER 3   *Leading Teams*

The Headline Test

Being a leader can encompass so many things, but the fact is that we all need to be reminded that "a leader of one can become a leader of many, but if you can't lead one, you'll never lead any." When people focus on solving problems instead of complaining about everything that is wrong, they are on the right track.

Making good decisions can be best accomplished by reminding yourself and your team members about the *headline test* (fig. 3–12). The headline test is simple. Take a few minutes to imagine how your actions would be seen if reported in the papers or on TV. In other words, how would your actions look on the front page of tomorrow's newspaper? What would the headline read?

Fig. 3–12. Act as if your actions were going to be on the front page of tomorrow's newspaper.

The Power of Words

Growing up in New Jersey, playing stickball in an empty parking lot was a typical Saturday ritual for most kids. Because everyone in New Jersey seemed to have a father, brother, uncle, or cousin in the fire service, it was not unusual for the kids to make a quick visit to the local fire station after the game. One kid in particular looked forward to talking with the firefighters in front of the fire station even more than he enjoyed playing stickball. By the age of 14, he was on a first-name basis with all the firefighters in town. When he stopped by to talk to them, he would ask them to share stories about fires they had fought. He listened intently as they told their tales, even if some of them seemed to get grander in scale over time (wink). Before he reached the age of 15, he knew there was only one career choice for him. He was going to be a firefighter.

At the age of 21, he realized his dream. After a few years on the job, he was assigned to the rescue squad on a busy department in northern New Jersey. He loved being a firefighter and took it very seriously. On one cold Saturday morning in January, his squad arrived first on the scene at a working structure fire in a single-family dwelling. He and his partner entered the home and made their way up to the second floor. They crawled down the hall until they reached the door of the room that the fire was in. He entered the room alone, wearing state-of-the-art bunker gear that he had purchased himself. This gear enabled him to move deeper into the room, so he could perform a search for an unaccounted-for teenager. However, his partner was wearing department-issued pullup boots and had to remain at the door because the heat was too intense.

The fire began rolling across the ceiling. As he crawled through the room sweeping his tool across the floor, his worst nightmare happened. A member of the ladder company who did not know he was inside busted out a window from the outside. The room flashed over. After two failed attempts, he was finally able to dive out a window onto a porch roof, but not before suffering severe third-degree burns on his neck and hands.

He is one of only a few firefighters to have survived a flashover. While being treated at the burn center, the doctors determined that he would need skin grafting to reconstruct both of his hands. After months of treatments and surgeries, he was back home, recovering and contemplating his future. During this time, he wondered if he still had what it took to be a firefighter. He knew his hands and body would heal, but would he still have the courage to do the job? It was a valid question.

After months of pondering the question, he knew he had to try. The day he returned to work, he still did not know the answer, so he took a temporary position as a dispatcher. He loved being part of the team again, but dispatching and fighting fires were two different things. He felt incomplete. After much thought, he felt compelled to request a transfer back to the line. He wanted to fight fires again. However, after the experience he had just been through, the question he continuously pondered was, "Am I willing to risk my life for a complete stranger again?" It was another valid question, and one that could only be answered at his next working fire. That day came sooner than he expected.

When his engine pulled onto the scene, smoke and flames were billowing out the side windows. His instincts took over and he began to stretch a hoseline toward the entrance of the house. The heat intensified as he approached the front door. His self-doubt increased along with the heat. He cautiously entered the structure and began to advance the line, when suddenly he just stopped, frozen with fear. He wanted to continue, but his legs felt as if they were frozen in a block of ice. He could not do it. The fire had won. He was about to drop the hoseline and walk home when his captain came up behind him, leaned in, and started talking in his ear.

"It's okay. I'm here with you. We're going to be fine," the captain said.

He shook his head to say no, but the captain was not having it. He was a seasoned officer, and he knew what was going on inside the firefighter's head, so he began to reassure the firefighter.

"One step at a time. I'm right here with you," the captain encouraged. "I know it's hot, but you're going to be fine. You can do it."

Knowing that his captain believed in him was enough to give the firefighter the confidence he needed to push forward and get the job done. Together, they put the fire out, and he proved to himself that he still had the courage and skills to get the job done. From that day forward, he never questioned his ability to perform on the fireground. He even went on to become a high-ranking chief officer and one of the most influential members of the New Jersey state fire service.

"I believe in you"—four of the most powerful words you can ever say.

When was the last time you told one of your team members that you believed in him or her? Can you count on one hand how many times you encouraged someone by saying, "I believe in you"? Has anyone ever told you they believe in you? If so, you are actually one of the lucky ones. I once gave a leadership presentation to a couple hundred fire service professionals in San Antonio, Texas. I asked the question, "How many of you remember the first time someone said, 'I believe in you?'" Surprisingly, only a few hands went up. At first I thought no one heard the question, so I repeated it. Still, only a few hands were raised. My disbelief was more obvious than I wanted it to be. As the father of three boys, I am constantly telling my sons I believe in them. My father said it to me, and so did a few other influential people in my life. I looked out into an audience of grown men, some in their late 50s, and realized most of them never had anyone say those words to them.

This is unacceptable. If you happen fall into that category, let's fix this now. You are actively seeking a better way to achieve success. You have chosen to spend your time educating yourself by reading this book. First, I thank you for the gift of your time. Second, I want you to know I believe in you. Keep doing what you are doing. Continue on the path you are on. Jack Canfield talks about his Law of Probabilities, which says "the more things you try, the more likely one of them will work." For example, the more books you read, the more likely one of them will have an answer to a question that could solve a major problem you are facing. Maybe it will improve your finances, solve a health challenge, or help develop a highly functioning team.

"I believe in you"—four of the most powerful words you can ever say to a team member. Say them often.

Praise in public, correct in private. Don't ever publicly embarrass, correct, or admonish anyone. You'll create a mortal enemy for life.

—Denis Onieal

Communication

Firefighters often fail to realize that almost everything we do in the fire service is based on writing and presentation, both of which are forms of communication. Think about the last time you were in your superior officer's quarters trying to solve a problem. Consider the most recent interaction you had with a group of colleagues in the fire station. How about that written document you helped prepare last month or referred to during a training exercise? They are all forms of communication.

Effective and appropriate communication is critical at all times and at all levels in our profession. The efficiency of a team begins with the sharing of information. Teams that do not communicate are destined for failure (fig. 3–13).

What are we trying to accomplish? How are we going to get there? What is each individual team member's responsibility? These are the types of questions your team members need to ask and know the answer to. It is true that teams fail when one person wants to take all the credit, but it is also true that teams fail when they lack direction and clarity of purpose.

Step Up Your Teamwork

Fig. 3–13. Communication is essential for a team's success. *Courtesy:* Tony Perez.

On the fireground, every company must know what the other companies are doing. Without communication, someone can get seriously injured or worse. Some of the stories in this book illustrate that point. It does not matter if you are fighting a fire or working on an important project. When companies do not communicate with each other, they cannot coordinate their efforts. When they do not coordinate their efforts, the end result is often failure.

Communication before delegation

If you intend to successfully lead a team, you must learn to communicate your vision with extreme clarity. Passing responsibility and delegating to others who are uninformed is a recipe for disaster. People need to *know* before they *go*. If you fail to communicate with your team members, they will be unsure what their individual responsibilities are. When you communicate clearly, you can avoid misunderstandings and prevent many unnecessary problems and mishaps.

You must learn to communicate your vision with extreme clarity. People need to know before they go.

Mistakes ineffective communicators make

Poor and inadequate people skills aside, there are three major mistakes that ineffective communicators often make.

1. Being unclear about what needs to be accomplished. You must know what you want your team to think, feel, and/or do. This is only possible if you know the goal you are trying to achieve.

2. Not preparing properly. Those who excel know the importance of taking time to map out a game plan and prepare for success.

3. Being unable to make people care about your message. If people do not care, they do not listen. One of the keys to solving this problem is to address how everyone benefits if the desired results are achieved.

Even in ideal conditions, you can have problems communicating. Do not try to be perfect. Remember, teams that communicate openly and honestly have a better chance of achieving success than teams that struggle in this area. When speaking to a team or trying to stress an important point, one way to ensure you convey a strong message is to use stories.

The Power of Stories

There is a popular slogan in the sales profession—Facts Tell, Stories Sell—that encourages salespeople to use stories to help move a product or service. Some may argue that it is nothing but a theory, but the fact is, it works. Our brains are programmed for stories. As humans, when we hear or read a story, our brain sees images. We imagine what people look like, what they are wearing, how their bodies are moving, and the locations where the stories take place. This helps us remember the point of the story.

The best communicators I have ever met were great storytellers. Skillfully told stories are one of the most powerful weapons in your teaching and communication arsenal. When you are communicating to your team or giving a presentation, incorporate stories into your talk, and you will see firsthand the effect they can have.

You can tell stories that share your personal experiences or the experiences of others to help make your point. In our field, stories that demonstrate close calls on the fireground are powerful because they make it easy for firefighters to mentally put themselves into the same situations. Some experts believe there is no better way to make a human connection and communicate than through a story. I happen to agree with them.

> *Skillfully told stories are one of the most powerful weapons in a leader's teaching and communication arsenal.*

If you need a little more convincing, here are eight reasons why you should use stories as an effective way to relay a message.

Eight reasons why stories aid effective communication

1. Stories prevent confusion. The mind does not easily absorb a PowerPoint slide with 19 bullet points (fig. 3–14). What is the point you are trying to make? It is easy to get distracted from your main objective when you are communicating and presenting, but stories have a beginning, middle, and end. A good story will have a message attached to it, which will help prevent confusion.

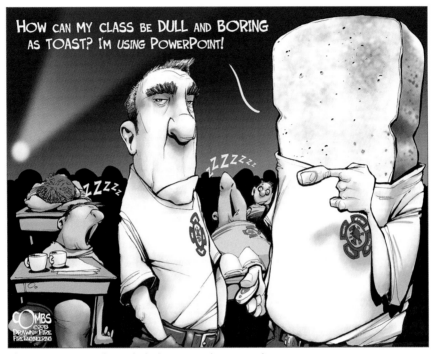

Fig. 3–14. Use stories to help keep people engaged.

2. Stories create curiosity. In a straightforward presentation, most instructors cut right to the chase. They explain a tool, what it is used for, how to use it, where and when to use it, and why it should be used. All of this is important, but dry facts are boring. When you tell a story, however, you bring people in and slowly lift the veil on how the tool can make their jobs easier. Curiosity engages the listeners, which will help them absorb the message.

3. Stories bypass skepticism. The natural condition of most people when they are presented with a new tool, procedure, or method is resistance. Most people in the workplace dislike change. Because of this, their minds are often closed. They do not want to change, and they do not want to look foolish; therefore, the easier thing to do is resist. But when you teach with a powerful story, there is little to resist against. You are not telling people what to think. You are simply showing them what happened in a similar situation to theirs, and leaving it up to them to draw their own conclusions.

4. Stories help you make a personal connection. People want to be led by someone they know, like, and trust. When you skillfully tell your story, mistakes included, others will feel more connected to you. When they see and hear how you overcame a challenge, they sense that you genuinely care about helping them improve and overcome challenges of their own.

5. Stories are believable. When you tell a good story, people visualize the setting and scenario. If I tell you the story about when firefighter Smith was climbing up an unsecured ladder to the roof of a two-story home, and the ladder kicked out a few inches, causing the firefighter to drop the chain saw, your imagination becomes active. You can see that scene in your mind's eye. There is a part of the subconscious mind that cannot distinguish from a true or imagined experience. This is why people get emotional at movies or when they read a good book.

6. Stories are emotionally suggestive. We all like to think that we make decisions based on facts, but human action is the product of emotion. If your objective is to get other people to take action, you need to get them emotionally connected. A well-presented story is the quickest way to make that connection. A person who is emotionally linked to a story will remember it for a very long time, possibly for the duration of his or her career.

7. Stories help you say things without offending others. When you criticize others who are doing things incorrectly, they become offended. When you tell a story about how you (or someone else) were doing similar activities, and achieved better results once you changed your actions or behavior, people have a tendency to be more open-minded. If you make yourself (the former wrongdoer) the key character in the story, people will not feel you are criticizing them but instead are educating them from your own personal experience.

8. Stories differentiate you from others. When you use great stories that resonate with the people you are speaking to, you stand out from the crowd. Your message will not be robotic; your communication and presentation style will differentiate you from the crowd.

As with anything else, storytelling as an educational tool takes practice, but it is worth the effort. When you develop the ability to tell stories, you increase your effectiveness as a communicator and your influence as a leader.

Public Speaking

Public speaking in general can be a positive and life-altering experience. When you stand strong in front of a room full of your peers, this translates into confidence. Success and confidence in one area of your life often carry over to other areas. This is because successful actions can act as a blueprint when they are repeated. You may not like the idea of speaking in front of others, but if you plan to lead a team, at some point you will have to address (at the very least) your team.

I will never forget the first time I gave a 45-minute business presentation to more than 100 people. I was so intimidated by the thought of public speaking that I read Dale Carnegie's book, *The Quick and Easy Way to Effective Speaking*, and took 13 pages of notes. Then I organized my speech a month before I needed to give it and revised it countless times. I gave a mock presentation 30 days in a row to an empty couch in my living room (similar to what I did when studying for my deputy chief promotional exam).

The day I gave the actual presentation, I never had to look at my notes. Some people said I was a natural, not realizing how diligently I had prepared. A few years later, I began speaking in front of groups, business teams, and associations, ranging from 10 to 5,000 people. Did I get nervous? Yes, and I still get nervous every time I speak publicly. If you do not get nervous when you speak in public, I am impressed and a bit jealous.

The best communicators use their own creativity to add the essence of who they are and what they stand for, which is why the outcome is always a little different for everyone. One of the best tips I could give you for public speaking is to search the Internet for videos from your favorite communicators and study them. Take notice of the moments when they capture your full attention. Listen and watch how they communicate and interact with the other people in the room. Do not watch with the intention of being entertained; watch with the intention of learning.

Step Up Your Teamwork

If you want to develop your ability to connect with a team, do not pass up opportunities to speak in front of people. You may not be comfortable at first, but in time, you will find your voice (fig. 3–15).

Fig. 3–15. It takes time for public speakers to find their voice. If you want to develop the ability to impact an audience, never pass up the opportunity to speak in front of a group.

The Attention Span of a Goldfish

We live in the age of commercials and technology. We are overstimulated. We have so much information coming at us that we need to filter out the garbage and retain the important. The BBC released a report stating that the addictive nature of Internet browsing can leave a person with an attention span of nine seconds—the same as a goldfish.

What this means to a speaker is if you cannot connect with someone in the first nine seconds, the odds are high that they are going to redirect their attention to something or someone else. You may be able to capture their attention later, but we also live in a time of cultural ADD (attention deficit disorder). There is an overload of distractions in today's society. Information used to be power, but not anymore. Now your ability to "surf the net" is power. The discipline once required to attain information has been replaced with the use of a cell phone and a search engine. However, be careful—the world is full of self-proclaimed experts who are willing to blog about anything. The difficult task now is separating the quality information from junk. Although we live in a high-tech society, it will be your ability to connect with people that will differentiate you from everyone else.

High Tech vs. High Touch

For you to succeed as a communicator, you will need to be not just heard, but understood. My wife and I can communicate about things without words. When our 10-year-old son was trying to muster enough courage to do a high dive into a pool, my wife looked at me and raised her eyebrows in a way that said, "I'm not sure about this." I answered by putting my hands out, palms down, and moving them downward in a way that said, "Relax, it's going to be okay." She then verbally replied, "Are you sure?" Our son, who was oblivious to our unspoken conversation, replied, "Am I sure about what?"

That is an example of how we communicate without words. But in order to connect with people when speaking in public, you will want to communicate with properly chosen words. There are two main reasons why words are important:

- Words are the primary way we receive information.
- Words are the primary way we give information.

Words have been spoken to us in the womb, before we were even born. Words are part of our psyche. Most people can finish the following sentences; "Give me liberty or give me . . ." or, "Ask not what your country can do for you. Ask . . ." Words have the power to hurt or heal, to encourage or discourage. They can have a positive or negative effect on a team. You can use words to inspire productivity or to destroy morale.

In today's high-tech world, all the same rules about communication and words apply. The best way to communicate is still face to face, but this is not always the way we do it. People use text messages and e-mails, and by the time this book is five years old, there will be some other type of electronic communication. This is fine, but it is important to know that words in print often do not translate the way they do in person. I once sent the following sentence to an officer after battling a tough fire: "I can't believe you put that fire out without a backup line." He thought I was reprimanding him for not waiting for the second engine company. I was actually complimenting him for making a great stop, especially since all our other resources were committed to other incidents at the time.

High tech is great, but these forms of communication should not replace *high touch*, which is not just communicating but connecting with people. It is still the best way to get through to people. My guess is that it always will be. This is why learning how to properly communicate with others is a valuable skill that will help you achieve a higher level of team success.

Reference

1. Colin Powell, "The Essence of Leadership" (February 10, 2011), http://www.youtube.com/watch?v=ocSw1m30UBI.

PREVENTING TEAM COLLAPSE

4

Complacency kills! If I could post two words on a wall in every fire station in the world, those would be the words. Complacency is the feeling of being satisfied with how things are and not wanting to try to improve. In corporate America, athletics, the military, and anywhere else where you find groups of people trying to enhance their ability to pull together, complacency will cause those teams to collapse. The fire service is no different; complacency is absolutely unacceptable (fig. 4–1).

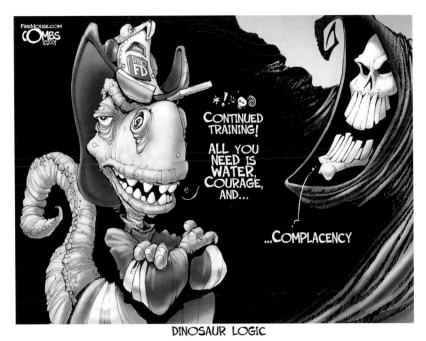

Fig. 4–1. Complacency is unacceptable.

This is especially true when it comes to how we train for and operate on the fireground. Throughout my career, I have heard firefighters refer to jobs as "routine fires." I cringe every time I hear those words. We cannot afford to pull onto a fire scene, or any incident for that matter, with the words *complacent* or *routine* on our minds. The incident that happened to my friend Mike early in his career explains why this is dangerous.

"Routine Fires" Don't Exist

Mike was the captain on an active engine company in Jersey City. He and his crew had just finished extinguishing a car fire when one of them noticed smoke coming from down the street. He approached Mike and said, "Cap, what's all that smoke from?"

It was a heavy column of smoke coming from a row of three-story mixed occupancies. The buildings were residential occupancies over stores located in a busy area called the Five Corners.

Mike keyed his radio microphone. "Engine 7 to dispatch, do you have something coming in for Newark Avenue?" he asked.

"Yeah, we're transmitting a box now, Cap. Report of a building fire," the dispatcher responded.

"We're wrapping up here. We'll take it," Mike replied.

"Received, Engine 7. You'll be first due," the dispatcher acknowledged.

The crew jumped back on the engine and rolled toward the fire. Upon arrival, Mike sized up the three-story ordinary brick building. The front was clear, but smoke and fire were showing out of two third-story rear windows. He confirmed the fire so dispatch could relay the message to the other incoming units.

Mike and a probationary firefighter named John began to stretch a hoseline in through the front door and up the common stairwell to the fire floor. The driver communicated with the second due engine crew, which was five minutes away. That crew would help secure a water supply from a nearby hydrant. A fourth firefighter helped stretch the line to free it from kinks and prevent the hose from getting caught up on corners.

Chapter 4 — Preventing Team Collapse

As they stretched a dry line up the stairs, Mike turned to John and confidentially said, "This is going to be a piece of cake—routine."

At the top floor was a light smoke condition. This gave Mike even more confidence because it was proof that the fire was venting out the rear windows. He believed the heat and smoke would be minimal, so he held off on charging the line to make the stretch easier. After years of experience with similar fires, he knew they would only have to briefly hit the fire with the hose stream to extinguish it. The other companies were heading to the scene. The engine company would have to work alone for the initial stretch and attack. This did not concern Mike. He had been here before, many times.

Mike and John began to make their way down a narrow hallway toward the rear of the apartment. They passed through a series of swinging doors that were between them and the room that was burning. The hinges on each door enabled them to swing open in either direction. The heat was minimal. They moved on their knees, but with their heads up high because the doors helped confine the fire and elements to one room in the rear of the structure.

When they reached the final door, Mike opened it and discovered the raging inferno. The room on the other side was fully involved, and the heat quickly banked down on top of them. Both men expected this, so neither found it alarming. Mike pulled on John's collar and directed him to back out of the room and into the hallway so they could charge the line with water and re-enter the room ready to attack. Once the door between them and the fire was closed, Mike keyed his radio and called for his driver to charge the line.

"Received, Cap. I'm charging the line now," the driver acknowledged.

The dry hoseline was snaked through the structure. They braced themselves for the water because the sudden surge would surely push the line out in awkward positions. This is typical, and usually only requires a few seconds to adjust before initiating the attack. Both firefighters sat in the narrow hallway just outside of the fire room and waited for the thrust of water. The fire roared as they impatiently held their position.

They continued waiting, but the water did not come. The volume of fire was so large and intense that it began to burn through the door in front of them. Their anxiety increased with the fire. The hallway continued to heat up. *Where's the water?* Mike wondered.

Finally, something began moving inside the hose. They braced themselves for the surge and waited with the nozzle aimed at the door. Still, nothing happened. The fire was now burning fast, and the heat began banking down hard on top of them.

"Open the nozzle," Mike called to John, anticipating that the water would fill the hose at any moment.

John pulled back on the handle and fully opened the line so water would flow freely. Instead, the water trickled out the end of the nozzle and onto the floor about two feet in front of them. The fire began to move into the hallway where they were staged.

"Increase the pressure. Increase the pressure!" Mike yelled into his radio microphone with growing impatience.

"You've got a full charge," the driver replied.

Looking at the pump panel on the engine, and judging from the solid hoseline that was heading into the structure, the driver was right. He was giving the exact pounds per square inch (psi) of pressure he was supposed to for the length and diameter of the hoseline they were using. Still, Mike was barely getting a drop from the nozzle. It resembled a stream from a garden hose more than a fire hose. The advantage was heavily in favor of the fire.

The fire began pushing Mike and John back into a wall behind them. Smoke filled the space, and they lost visibility completely. Mike made the right decision to back down the hall and wait for water at a safer distance. He reached behind him and felt a wall on his right and on his left.

Where's the door? he wondered. He was slightly disoriented and began feeling around for the door. It must have closed behind them. He swept the wall with his gloved hand, feeling for a doorknob. No success. *Where the heck are we?* he thought.

Relying on his training, he reached down and grabbed the hose so he could follow it out of the room. Surely the hoseline would lead to the open doorway. It was at that moment he discovered that the swinging door had closed on the hose before it was charged. When the driver sent water through the hoseline, it was pinched under the swinging door. The bottom of the doorframe acted as a clamp, stopping the water from passing through. But it was worse than that. Now the door was jammed in the closed position, preventing Mike and John from retreating. They were trapped in the burning hallway without water. It was a worst-case scenario.

Using every bit of muscle he had, Mike tried to push the door open, but it did not budge. The fire was rapidly growing. The intense flames pushed John back onto him. Mike was stuck between the firefighter and the door. Then it occurred to him that the door swung both ways. If he could not push it, he might be able to pull it. This was easier said than done. To pull the door, he would have to get John off him first. The flames were only a few feet in front of them. Between the heat and the weight of John on top of him, Mike was helpless. He could no longer reach his radio to call for help. The two of them were burning up. It was so intense that John began screaming from the heat.

"Don't stop flowing water!" Mike screamed back at him.

Although they only had a small amount of water flowing through the pinched hoseline, Mike knew they would need every drop they could get on the fire.

"Point it up toward the ceiling!" Mike yelled, hoping to keep the flames away from them.

While this was happening, the second engine company had arrived on scene. The members of that company were on their way down the hallway with a backup line. They could hear Mike and John screaming. They noticed that the first swinging door was closed on the line. This meant that not one but two doors were pinching the line. They forced the door open and off of the line and moved swiftly down the hall toward the distressed firefighters. They came up to the second door and could hear the trapped firefighters on the other side. They attempted to push the door open but encountered resistance from Mike and John, who were pinned back onto the door. The charged line was still wedged underneath, preventing them from moving the door in either direction.

This is it, Mike thought. *We're done.*

The fire was rolling above their heads. It felt as if the flames were touching their facepieces and wrapping around their helmets. Their air was running low, and they were exhausted. The room was seconds away from flashing over.

"I'm burning up!" John screamed.

"Keep it flowing. Don't shut that line!" Mike screamed back.

The captain could hear their conversation from the other side of the door. He knew they were pinned down. He unsuccessfully tried pulling the door open. Then he pushed against the door with everything he had. Mike and John resisted, but the captain knew this was the only way he could free them. If this did not work, it was going to be a very dark day for their department.

The captain pushed hard. It was just enough to get the door to move, freeing the hose from beneath. Mike and John were still pushing in the opposite direction, but the captain was able to open the door enough for the trapped firefighters to fall back into the hallway, where the second engine crew was positioned. The two pushed their way past the others and raced down the hall and down the stairs as fast as they could, fearful of running out of air. The fire had gained so much momentum that it pushed the second engine company back and out of the building as well. When all of them reached the bottom of the staircase, the incident commander pulled everyone out of the building and switched to a defensive attack.

Understandably, the incident had a big impact on Mike. Over the next few days, he analyzed what happened and wondered what he could have done to prevent it. He could not get over the fact that he put himself and his team member in that position. He made a promise that it would never happen again. After the incident, Mike became a walking hardware store. His turnout coat pockets were full of wedge-shaped pieces of wood that he used to chock every door he walked through to prevent it from closing behind him. He preached the importance of situational awareness to any firefighter who would listen. He promised never to advance an uncharged hoseline down a hallway toward a fire again. He also swore never to refer to any job as "routine" again.

Incidents like that should make the hair stand up on the back of your neck. That is the effect it had on me when I first heard the story. It serves as a reminder that if something can go wrong, it usually does. There is no such thing as a "routine fire." Those words should not exist in our vocabulary. To be fair, there is one time when you can legitimately consider referring to a fire as routine—after the job is complete. Until then, firefighters must do everything in their power to prepare for the unlimited number of challenges that often accompany structure fires, because complacency kills.

Complacency may not kill someone on your team, but it absolutely can cause any team to collapse. Think about areas where your team has become complacent. What situations have become routine in your daily operation? The time to take corrective action is now, not after the tragedy strikes.

There is no such thing as a "routine fire."

Why Teams Fail

A fire director called a group of officers into his office to discuss something important. The department had dwindled in size from 180 members to 125. Companies had been placed out of service indefinitely, and the city had been talking about possible layoffs. To make matters worse, 7 firefighters had been injured at the last 10 multiple alarm fires, and town officials did not seem to care.

The officers were tired of being treated with no respect. They were forbidden to speak in public about the situation, under the threat of disciplinary charges for conduct unbecoming an officer. Their hands were tied. When the fire director called them to the meeting, the officers were expecting him to explain that he was as frustrated by the situation as the rest of them. Hope was on the horizon, and the officers were excited that things were about to change.

Unfortunately, they could not have been more wrong. The fire director was more interested in chastising them for things like misspelled words in reports, lack of participation at a recent community ceremony, and decreasing morale around the firehouse. He told the officers that they were losing control of their groups and the department. The director stormed out of the room before the officers had a chance to comment or share their views about what was happening. They sat in the room, bewildered, wondering if the meeting was over. One of the senior members stood up and said, "Well, I was thinking about retiring. At least he helped me make my decision. I'm done with this crap." And he walked out.

That story illustrates a team that has collapsed mainly because their leader is out of touch with the rank and file. Of course morale was low among new members; they were in danger of being let go by a city dealing with a budget crisis. This type of disconnect between upper and middle management happens in every industry on a regular basis. Team failure, however, is not always about one person being out of touch. There are many reasons why teams fail. For example, teams fail when one person wants to take all the credit. My answer to that is if you want to take all the credit when things go right, you have to be willing to accept all the blame when things go wrong. Teams also fail when they are consumed with drama. Drama in the workplace occurs when a team is not focused on a goal. A team on a purposeful mission does not have time for drama. Below are eight of the top reasons why teams in the fire service fail. Take a quick read and see if one or more apply to your team.

Eight reasons why fire service teams fail

1. Poor communication. Without strong communication, a team simply cannot function. Someone has to be able to communicate the purpose of your team's existence. People need to have a game plan. Without direction, you are leaving it all up to chance. Team members need to be well-trained, and they need feedback about what they are doing well and what they need to improve. It is not sufficient for only the leader to know what he or she wants to accomplish; team members need to be informed also. They need to know what the mission is, and the time frame and tools they have available to get the job done. They need to know what is acceptable and what is not. Relationships fail when two people do not communicate. Teams are no different (fig. 4–2).

2. No vision. Throughout this book, the words *goal* and *vision* are used. This is not by accident. A *vision* is a mental image of what your team is trying to accomplish. That is where the concept of a vision statement comes from (similar to a mission statement). Having vision means you see your team achieving success in the future. One or two team members may set the standard and ultimately determine the end goal, but everyone on the team should be included in establishing the vision. If your team is operating without clarity, they will not work cohesively. When this happens, members lose enthusiasm and motivation. Your goals may change along the way, but remember: a team without vision is destined to fail.

Fig. 4–2. Communication is necessary in order to achieve success as a team.

3. Avoidance of accountability. Lack of accountability is not only a problem on the fireground. Have you ever worked on a team where it seemed no one wanted to work or take responsibility? That is an enormous problem. Teams are unlikely to achieve success if even one person refuses to fulfill his or her role. When a person fails to be personally responsible, that person also fails to be accountable to others on the team. The success of any team is largely dependent upon the involvement of every person on the team. Bringing talent, skill, and ability to the table is only half the equation. The other, more important part is utilizing those skills to better the team.

Sometimes lack of accountability can be the result of undefined roles. Sending a group of firefighters into battle without assigning them specific tasks will only cause confusion for everyone. A ladder company sent to the roof to ventilate has a completely different task from the engine crew stretching the initial attack line. Both groups are on the same team, trying to accomplish the same goal, but if their tasks are not defined, the mission is doomed. The same can be said for business teams, sports teams, and recreational teams.

4. Unresolved conflict. Put five or more people in a room, and eventually there will be friction. Even if they appear to be best friends, in time they will be forming alliances and becoming increasing hostile toward each other. Try to name 10 rock bands that have stayed together (with their original members) for 10 years or more. It is nearly impossible for one simple reason—egos. Egos often lead to unresolved conflict. We are all different. I cannot stress enough the importance of identifying and embracing the uniqueness of each member of your team. But the key is for the rest of the team to understand that your differences are what enable you to become stronger (fig. 4–3). If this is not understood, those same unique qualities are also the very things that cause major conflicts. When tension occurs on your team, it must be identified early and dealt with in a timely fashion. Failure to do that will cause the problem to intensify. A small, unresolved conflict among two or more members can become a runaway freight train that will dismantle the team and result in failure. Conflict is inevitable, but if handled properly, it can help you develop a stronger team. (Tips for conflict resolution are provided later in this section.)

5. Power struggles. A team leader with an out-of-control ego will destroy a successful team in record time. The only thing worse than a leader with an inflated ego is two leaders with inflated egos. In the fire service, we often joke that the term CHAOS stands for Chief Has Arrived On Scene. It's not uncommon for a group of firefighters to feel that everything was going well until the department head decided to show up. This, of course, is only a problem when that individual is a micromanager who thinks his or her ideas and methods are better than everyone else's. According to statistics, there is a 70% chance you have worked for a micromanager at some point in your career. If you are contemplating whether this paragraph applies to your personality and leadership style, let me help you out. It probably does. The good news is that you can fix it. Begin by realizing that power struggles make an entire team uncomfortable. When two people are jockeying for position as the leader, the rest of the team becomes detached and eventually inactive. The end result is a dysfunctional team that will not have the ability to accomplish even the simplest goals.

Fig. 4–3. Embrace the uniqueness of your team members. Your differences are what will enable you to become stronger. *Courtesy:* Tony Perez.

According to statistics, there is a 70% chance you have worked for a micromanager at some point in your career.

6. No clear identity. This is not usually a problem within the fire service. We understand why we exist. Our mission is clear: life safety, incident stabilization, property conservation, protect the weak, and serve the public. There is no mistaking what a fire organization does. We have had books written about us. Many movies and television shows have revolved around people doing our job. We do not have to advertise. When people need us, they call 9-1-1, and we appear. We know why we are here, and so does the rest of the world. With that said, sometimes teams lose their way. Does your team have a clear understanding of its purpose and direction? A strong team will know its direction, focus, and goals. A strong team will take pride in serving its customers and being positive role models for our children. It is important that your team clearly understands who it is and why it exists.

7. Poor training habits. Let me put it in clear terms: when a fire department spends more time on the fireground than on the training ground, it is destined to fail (fig. 4–4). This same theory applies to any performance-based organization. You can never train too much, but you can certainly train too little. Worse yet, your team can be spending sufficient time on the training ground, but doing all the wrong things. Practice will not make your team perfect. Perfect practice will make your team perfect. We have no shortage of true experts in our field. If your team is not getting the results it wants, bring someone in to help you get on track. Every organization in America brings in an occasional expert to educate and stimulate the team. Mistakes are common in this world, but the time to learn is not when the alarm comes in. We need to be on top of our game long before the incident. Make it your goal to learn something new every day.

8. Improper strategy and tactics. This falls in line with the last point but takes it one step further. Improper training will lead to bad decision making. Strategically, the first few minutes are the most important part of the fire. That is when we position apparatus, stretch lines, ventilate, and put our plan into action. That is also the most stressful part of the incident, which means we have to make incredibly important split-second decisions. The ability to do this comes from preparation. Spend time in the books, on the training ground, and at the seminars. Train yourself and your team to make smart decisions and fewer mistakes.

When you make mistakes, learn from them so they are not made again. A mistake made twice is no longer a mistake; it's a choice.

Fig. 4–4. Successful teams spend more time on the training ground than on the fireground. *Courtesy:* Tony Perez.

Other reasons teams can fail include poor dynamics, absence of trust, poor time management, unclear purpose, lack of fun, and inattention to results. If you see any of the signs outlined in this section, be proactive and take corrective action before your team collapses.

A mistake made twice is no longer a mistake; it's a choice.

Dysfunctional Teams

Truss roofs are one of the greatest challenges for firefighters when it comes to structure fires. Large open floor plans such as bowling alleys, car dealerships, warehouses, and movie theaters are just a few of the types of structures often constructed with truss roof components. In more recent years, residential building contractors begin using truss-like configurations in lightweight constructed buildings. When sections of these roofs are exposed to fire and heat for an extended period of time, failure is inevitable. The challenge we face when this occurs is that rarely does only a small section of the roof fail. Instead, when one section fails, the entire roof is prone to collapse. Such is the case with a dysfunctional team.

In Patrick Lencioni's book *The Five Dysfunctions of a Team: A Leadership Fable* (San Francisco: Jossey-Bass, 2002), he tells a story of a corporate CEO whose goal is to unite and build a high-performance team. In the book he identifies the five dysfunctions as follows:

1. Absence of trust
2. Fear of conflict
3. Lack of commitment
4. Avoidance of accountability
5. Inattention to results

They are all valid and mostly self-explanatory. However, in the fire service we have identified three additional "common" dysfunctions. They are as follows:

1. Poor communication
2. Low morale
3. Freelancing

When one or more of these dysfunctions are present in your team, they must be addressed. If not, it is only a matter of time before one dysfunction leads to another, and you will be in jeopardy of total and complete collapse. To illustrate the point, take note of the challenges that plagued one officer's team.

Chapter 4 — Preventing Team Collapse

Captain Bryant had his hands full. His crew had been arguing among themselves for weeks. It had originally started as a disagreement between two members. Firefighter Smith complained that Firefighter Reilly was lazy around the firehouse and dangerous on the fireground. The discussion became heated enough that the others had to separate the two members. As time passed, Firefighter Reilly began to regularly show up late for work, distancing himself from the others and avoiding his daily housework. Animosity grew throughout the team as the others became bitter toward Reilly, who was the only team member they felt was not pulling his weight. They also felt it was unfair that Reilly was getting away with doing less than the others were required to do, which lowered morale.

Captain Bryant tried to talk with Reilly several times, but the firefighter seemed distant and rarely gave more than one- or two-word answers. Smith felt that their captain wasn't handling the situation properly, so he requested a transfer to another shift. As Captain Bryant was contemplating the situation, a call came in for a working basement fire. Reilly was late getting on the rig. When his engine company arrived on the scene, the captain told Smith and Reilly to stretch a line and follow him down the interior stairs to the location of the fire. Not able to see through the dark smoke, the captain lost contact with Reilly and was unable to reach him via radio. He pulled out of the basement and began to search for his missing firefighter. After calling a Mayday and requesting the rapid intervention team to help find his missing crew member, the captain was informed that Reilly had been located outside the structure.

The angry Captain Bryant approached the firefighter and asked where he had been. Reilly replied that he had lost contact with the other two and his radio was not working, so he conducted a search of the first floor alone until his air bottle ran low. When his breathing apparatus alarm activated, he retreated to the exterior to regroup with his company. Captain Bryant noticed that Reilly had plenty of air left in his breathing apparatus. When a radio test was conducted, it worked fine. The next working day, Reilly called in sick and three others requested transfers off the shift. They told their captain that they did not care about doing their job the way they used to because there seemed to be a double standard.

In that example, all eight dysfunctions occurred to some degree. The firefighters did not *trust* Reilly. Captain Bryant seemed to *fear the conflict* that he had to face in order to properly address the situation. His inability to deal with the issues caused his team to question his leadership ability. Reilly showed a *lack of commitment* to the team and their mission. He also *avoided accountability* and displayed a *negative attitude regarding the results* they were trying to attain. *Poor communication* efforts were evident and needed to be improved between all parties. Reilly's *freelancing* on the fireground was dangerous and could have resulted in the injury or death of his coworkers. Together this resulted in a *decrease in morale*.

When it comes time for this team to perform at a high level, what do you think their chances of accomplishing a challenging task will be?

Some organizations don't need competitors; they have themselves.

—Frank Viscuso

Team BLEVE

BLEVE is a fire service acronym for a boiling liquid expanding vapor explosion. A BLEVE is explosion caused by the rupture of a vessel containing a pressurized liquid heated above its boiling point. This type of explosion can easily occur when the container is exposed to flame and heat.

If dysfunctions are not dealt with soon enough, a team BLEVE can occur, such as in the previous story (fig. 4–5). Team leaders who find themselves in similar situations will need to conduct a formal subordinate interview like the one outlined in *Step Up and Lead*. Most of the time, however, a formal interview is not necessary, and some basic conflict resolution skills are sufficient. However the situation is handled, the team leader must address any of the eight dysfunctions observed with his team. Otherwise, minor situations can intensify until total team failure is imminent.

Fig. 4–5. Dysfunctions and unresolved conflict can cause a team to explode the same way a vessel containing pressurized liquid heated above its boiling point does. *Courtesy:* John Harrison.

Conflict Resolution

When people work together, conflict is a part of doing business. This is especially true in the fire service, where we might spend 24 or more straight hours in the same room with a handful of people. Conflict is normal and should be expected to a certain degree, but unresolved conflict is also one of the top reasons for low morale in the workplace. When morale is low, there will be signs of reduced work ethic. A loss of work ethic results in decreased productivity and increased absenteeism and sabotage. Some may even express a desire to quit altogether, which ultimately could mean a decrease in customer service and an increase in complaints. It has been estimated that managers spend at least 25% of their time resolving workplace conflicts. This causes a big decrease in workplace performance, because resolving conflict requires time and energy.

Handling and resolving conflict in the workplace is one of the biggest challenges team leaders and fire officers will encounter. There are many variables that play into the way we handle conflict—so many that if anyone tells you there is one way to handle all conflict regardless of the scenario and people involved, they are wrong. There are two common responses to dealing with conflict: avoidance or stubbornly battling it out. Both of these methods are incorrect and will leave all parties involved feeling dissatisfied because no resolution has been achieved. Trying to convince one party to the other party's point of view without compromise rarely works. An old proverb sums it up best: "A man convinced against his will is of the same opinion still."

> *A man convinced against his will is of the same opinion still.*
>
> —*Unknown*

Conflict resolution is a skill that must be developed. Everyone, regardless of their position of authority on a team, can be a manager of conflict. All you need is a general understanding of what conflict is and a basic knowledge of how to resolve it. People who are involved in conflict will find great benefits in developing their own ability to avoid or resolve situations before they escalate.

When an IC asks a firefighter to force entry through a steel door so the team can access the building and put out the fire, he or she expects the job will get done in a timely manner. The IC is not overly concerned with how the task gets accomplished as long as it gets done. Firefighters understand this. When firefighters train on vehicle stabilization and extrication techniques, they use various tools and methods until they find the most effective way to accomplish this task (fig. 4–6). The same could be said about conflict resolution techniques.

Team leaders must learn to constructively mitigate conflict in a productive way in which all parties feel they have been heard. *Compromise* is often the key word when it comes to conflict resolution. A good leader will use that word when talking to the people who are dealing with a potentially explosive situation.

Fig. 4–6. We must practice our conflict resolution skills the same way we practice techniques for fire suppression and vehicle extrications.

Sources of conflict

There are many causes of conflict, and some of the primary ones are listed here.

Poor communication. Also listed as one of the top reasons for dysfunctional teams, poor communication is the primary reason for conflict in the workplace. Different communication styles or failure to communicate at all can lead to misunderstandings between employees or between an employee and a manager. Overall, 10% of conflict comes from difference in opinion, and 90% comes from the wrong tone of voice. Poor communication also may occur when e-mail and text messages are misinterpreted, causing conflict and animosity between two or more people.

> *Overall, 10% of conflict comes from difference in opinion, and 90% comes from the wrong tone of voice.*

Difference in personalities. Everyone has a different background and experiences that have shaped their belief systems and personalities. When people cannot agree to disagree on certain matters (like religion or politics), conflicts tend to arise. Even if taboo topics are removed from the equation, when personalities clash, conflict follows. People who are direct and straightforward, for example, often offend those who do not respond well to that type of personality.

Different values. This is a very common cause for conflict in the firehouse because there is often a generational gap between coworkers. Often conflict does not result from the difference in values, but instead from failure to accept the differences. When this happens, people tend to treat this as an insult to their character, which causes dislike and sometimes hatred.

Poor performance. When individuals within a team are not pulling their weight and it is not addressed, a reduction in morale and feelings of resentment are inevitable. This leads to conflict. One recent fire service poll listed the top cause for low morale in our industry as having to support "dead weight," otherwise known as lazy coworkers no one would discipline.

Competition. Healthy workplace competition can be great, but unhealthy workplace competition can tear a team apart. When people try to constantly "one up" each other because their egos need the attention, the result is often unhealthy competition that discourages teamwork and promotes individualism.

Those are a few reasons for conflict in the workplace. There are also other sources, such as differing interests, scarce resources, and past history. The key is to recognize conflict when it arises and address it.

Addressing conflict

There are different methods that can be utilized to address conflict on your team, and some are better than others. The correct way to resolve a situation will depend upon personalities and the actual problem that has caused the conflict. Here are the most common methods that people use when confronted with conflict. When you see these methods in writing, you can better understand why some of these techniques do not work.

- **Force.** Using your position and power to demand a specific outcome.
- **Avoidance.** Ignoring the issue and hoping the conflict will go away.
- **Competition.** Letting them go head-to-head with the attitude "may the best person win."
- **Accommodation.** One person surrenders his or her own needs to please the other person.
- **Collaboration.** Working together to find a mutually beneficial solution.
- **Compromise.** Finding the middle ground so each party gives a little and gets a little.
- **Separation.** Although this resembles avoidance, there are times when this is the only option.

Most experts generally believe that collaboration and/or compromise are the most productive ways of addressing conflict. They are also the methods that promote team building because they involve working together to find the best possible solution. With either method, there will be no definitive loser, and there is a possibility that all parties involved can feel satisfied.

There will be times when one side wins out altogether. For example, if conflict arises between two people, and one person's point of view seems to be the right one, a leader can choose to side with that person. In this case, the person must understand the need to act on what is right and to follow through on the agreed upon position.

Conflict is not something that only happens in the workplace among coworkers. It can also be something that occurs between us and our customers—yes, even in the emergency services. One of the most common occurrences is when residents are asked to evacuate the area for their own safety. People often argue against our reasoning, which sometimes leads to an uncomfortable standoff (fig. 4–7). Regardless of where or when conflict occurs, it is in our best interests to resolve the issue as soon as it is identified.

Fig. 4–7. Conflict sometimes occurs between rescuers and potential victims.

The conflict resolution process

Arriving at a positive resolution is the goal. The following steps can be taken when attempting to resolve conflict issues that have gotten out of hand. When a situation escalates to an almost unmanageable level, you will have no choice but to address subordinate issues in a more formal and structured format. If you catch these issues early enough, the following eight steps may be all you need to put an end to the problem.

1. **Maintain confidentiality.** Choose the appropriate time and place to address issues face to face, in a private setting.

2. **Don't take it personally.** Focus on the situation, not on the individual.

3. **Identify the central issue(s).** Clearly address your concerns and any obvious causes of conflict.

4. **Describe the situation as a mutual problem.** Explain that resolving conflict will benefit more than one party.

5. **Allow each person to speak.** Encourage people to share their views so you can attempt to negotiate the differences.

6. **Work out a solution together.** Remember, the key word is compromise. Determine which solution is the best option.

7. **Implement a solution.** Inform all parties of the potential consequences of unresolved issues.

8. **Monitor the outcome.** Review, evaluate, and revise as necessary.

Here are some helpful tips to keep in mind during a conflict resolution meeting.

Take time out. If emotions interfere with a productive resolution, take a break and resume at another designated time. Do not force a solution that you may come to regret.

Use leading questions. When working together to develop a solution, it would be wise to skillfully choose your words. If you know that Joe is always complaining about the lack of efficiency around the workplace, you know that efficiency is Joe's trigger point. In this case, you could say, "Joe, I think if we do it this way, it would enable us to work more efficiently as a group. Wouldn't you agree?"

Words are powerful. Words have an impact, especially when they come from a person of authority such as a boss, officer, or supervisor. Too many people bark out orders and criticize when they are hot under the collar and end up regretting what they said. When you find your levelheadedness being tested by a difficult team member, don't fly off the handle. If it needs to be addressed immediately, implement the 10-second rule.

The 10-second rule. If someone explodes, confronts you, or challenges you in a way that you feel is inappropriate, think before you react. Take 10 seconds, enough time for two deep breaths, maintain your composure, and then respond accordingly. This will help you to remain calm and in control of your own emotions. It will also allow you to approach the situation with diplomacy. When dealing with conflict and confrontation, remind yourself that the person who stays in control always has the advantage.

Conflict and what is commonly known as "the storming stage" are par for the course with any team. Conflict is not always a bad thing. Sometimes it is a sign that people on your team are passionate. Conflict can be healthy as long as the team does not remain at odds. Different people and situations will require different methods of resolution, but escalating conflict must be dealt with. Avoiding conflict may be the easiest way to deal with it, but that will only give the problem a chance to take hold, intensify, and break down a team. By actively resolving conflict when it occurs, we can create a more positive work environment for everyone.

> *To resolve conflict before it ever takes form is the highest ideal.*
>
> *—Bob Burg*

Handling Difficult People

As a team leader, you should be aware of the different personality types that you will be working with, especially those who will inevitably present problems. Difficult people can do to a team what oxygen does to a structure fire—intensify the problems (fig. 4–8). In the fire service, officers who have management responsibilities are held to a higher standard than managers in most other arenas. Simply put, as a fire officer, you are not only responsible for the outcome at incidents in which your team has performed, you are also responsible for the actual actions of your individual team members. One of the biggest challenges

I faced when I was a newly appointed captain was when I was brought into my chief's office and expected to answer for the actions of one of my firefighters. This was especially difficult since I had inherited a firefighter with a problem personality who had 20 years seniority on me, and now I was expected to "fix" him. It did not take long for me to understand that attempting to change people who refuse to change is one of the greatest challenges. Certainly we can set rules, but we cannot make people care.

Fig. 4–8. Difficult people can do to a team what oxygen does to a structure fire—intensify the problems. *Courtesy:* Constantine Sypsomos.

My experiences in business, sports, and the fire service have brought me to the conclusion that 95% of the people we deal with are easy to work with. Only 5% are difficult. Unfortunately, those 5% are much more vocal than the other 95% and therefore require a great deal of time and energy. A colleague of mine once said, "They may only account for 5%, but those 5% are more committed to their cause!"

This small but vocal group will test your sanity. They are the reason why psychiatrists have job security; if you spend too much time around them, you suddenly feel the need to seek professional help. Undoubtedly you know the type I am describing because you may work with them yourself. Perhaps they speak to you in a condescending tone, or maybe they fail to recognize problems or accept input. Some may show disregard toward you and your team, while others may walk around stressed out and making everyone around them equally stressed. They may be complacent, disinterested, or negative. Perhaps they are rude and arrogant. Whatever behaviors they exhibit, learning to deal with them is crucial if you intend to prevent team collapse.

This section is intended to teach you how to recognize and handle difficult people. You will also be introduced to the nine classic problem personalities we tend to encounter within the fire service and learn how to cope with and motivate each type.

Every fire department "rule" has somebody's name on it. You'll be much more successful as a manager if you take on the problem person rather than making another rule for the 1,000 nonproblem people.

—Chief Denis Onieal

Affirm

One effective way to handle difficult people is similar to the way we would handle a difficult fire: by using sound strategy and tactics. Follow this simple six-step process of assess, feel, formulate, initiate, review, and modify. It is appropriate that the first letters of these six steps spell the word *affirm* because as a team leader sometimes you must gather facts and assert your position so you can begin to resolve the issue.

1. **Assess the situation.** Ask yourself the following questions: Does this person present an ongoing problem, or is this a first offense? What type of personality does the person have? Will he or she

respond to an open discussion, or will you have to approach the situation more subtly? Are the actions of this person affecting the rest of the team in a negative way? Clearly define what the problem is and how it is affecting the chemistry of the team.

2. **Feel the person out.** If the complaint about one team member came from another, make sure you make an effort to experience or at least assess the potential problem firsthand. Taking action without sizing up the situation can be a bad idea and typically produces poor results. It is easier to tackle a building fire when you know what is burning.

3. **Formulate a strategy.** Difficult people do not just "go away," and it would be foolish to think you can simply change another person. You will need to come up with a strategy. The way you approach one difficult person may be completely different from the way you approach another, depending on their personalities and behavior types. Determining a strategy without knowing the type of person you are dealing with could be counterproductive, or at the very least, ineffective.

4. **Initiate a game plan.** Thinking about your strategy without implementing it will not accomplish anything. If you choose to do nothing about poor behavior or performance, you are allowing it to happen, and you will find it impossible to change. Being silent about a person's incorrect behavior is the same as giving permission to continue acting in that manner. Furthermore, failure to act will send a signal to the others that you are operating a team without rules or behavioral guidelines. This may have worked in the Wild West, but we have come a long way from the times of settling our differences with a shootout.

5. **Review your progress.** Once you initiate your game plan, keep a close eye on the outcome. Has the situation been resolved? If not, you may need more time. However, you might consider whether it is time to abandon the plan and come up with another strategy entirely.

6. **Modify your plan.** Once you have modified your plan, implement it. You will find that most of the problems you encounter when dealing with difficult people are resolvable, but make no mistake about it, some will not be.

Although this cookie-cutter format can help you handle difficult people, managing and leading a team is far from simple and routine. Each person on your team will respond differently to a specific communication style. A good leader will understand which methods of communication motivate each team member (fig. 4–9).

Fig. 4–9. Different personalities respond to different techniques and strategies. A good leader will understand which methods of communication motivate each team member. *Courtesy:* Constantine Sypsomos.

Nine classic problem personalities

Here is a list of some classic problem personalities (with colorful nicknames) and some tips on how to cope with and motivate each behavioral type.

The commander. Commanders are demanding and domineering. They are stereotypical control freaks who have a strong desire to prove to themselves and others that their point of view is right and yours is wrong. Commanders thrive on winning, especially in front of others. They can be bossy, rude, and have a short fuse (even though they do not see it that way). Commanders, like schoolyard bullies, can be abusive, overwhelming, and intimidating to others.

How to cope with the commander:

- Be assertive. They often respect leaders who stand their ground.
- Have them sit down, make eye contact, and expect them to give you their attention.
- State your opinions clearly but avoid saying things that will promote a fiery confrontation.

How to motivate the commander: Commanders tend to respond well to delegation, but you must be clear, direct, and concise about your expectations. Be cautious with the way you talk to them because they feel threatened when they think someone is attempting to take away their authority. Let commanders know how valuable they are, but do not fail to address how challenging their personality type could be for others on the team.

The piranha. Piranhas are like snipers. They do their dirty work behind the scenes. They deny their actions and avoid confrontation, but when things do not go their way, piranhas take shots at their prey from a distance. Piranhas are the type of people who criticize, condemn, and complain about others behind their backs (fig. 4–10). They are similar to the people who spread computer viruses or make anonymous calls to the press to spread negative information about an unsuspecting team member.

How to cope with the piranha:

-Bring them out of the shadows by raising issues in the open.
- Ask open-ended questions that require more than a yes or no answer.
- When you bring attention to the issues at hand, you inform them that you are aware of the situation, and they usually stop.

How to motivate the piranha: To get good performance out of piranhas, let them know that you have noticed the good job they have been doing and ask them to get more involved with the team. If you let them know they are on your radar, you will have a better chance of channeling their energy in a positive manner.

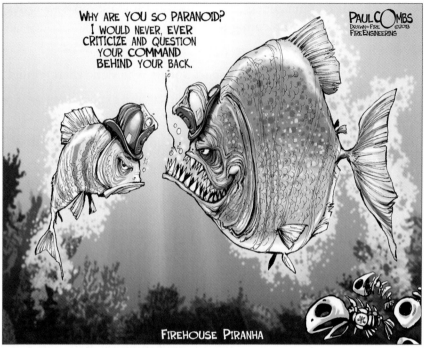

Fig. 4–10. Piranhas talk behind others' backs.

The wall puncher. Wall punchers are the most dangerous behavioral type. They are hostile, aggressive, and wired like a time bomb waiting to explode. They are capable of attacking a person with uncontrollable rage. To provoke a wall puncher is to virtually guarantee an intense confrontation. Wall punchers interpret any critical feedback as a sign of disrespect.

How to cope with the wall puncher:

- Approach wall punchers in a nonthreatening way, with your hands moving downward, as if to say, "relax."
- Call out the person's name until you break the tantrum.
- Be patient and give them time to regain their control before attempting to talk sense into them.

How to motivate the wall puncher: Begin by asking "self-convicting" questions such as, "What do you believe to be the most important characteristics of team building?" or "In your opinion, what is the quality of interaction among the members of our team?" You are leading up to the most important question, which is, "What can you do to positively impact the team?" Follow up their answers by praising them for being the type of people who are willing to do the ugly, unpopular jobs that no one else wants to do.

The compulsive pleaser. The compulsive pleaser is often called a *yes-man*. People with this personality type want to please everyone and therefore cannot make quick, accurate decisions. They need to be accepted, so they always tell you what you want to hear. They will make you believe they are in agreement with your plans, only to let you down in the end because they are not really contributing.

How to cope with the compulsive pleaser:

- Give them three choices and ask them to choose one.
- Encourage them to share their own thoughts on the subject or problem at hand.
- Do not allow them to make unrealistic commitments.

How to motivate the compulsive pleaser: Giving feedback to pleasers is done most effectively through the "sandwich technique," which involves inserting a criticism between two compliments. You begin and end the conversation with positive comments and praise, while ensuring that in between you adequately address the matter at hand.

The know-it-all. This type needs little explanation. They believe they know more than anyone else in the room, and they do not hesitate to share that belief. Knowledge is their bedrock, but a knowledgeable person with this type of attitude is about as useful as a 200-pound encyclopedia. It may contain all the information you need, but it is annoying and takes up too much space (fig. 4–11).

How to cope with the know-it-all:

- Encourage them to consider the other viewpoints.
- Be prepared to provide them with facts to back up your theories.

How to motivate the know-it-all: To utilize the strengths of a know-it-all, ask questions and listen intently to the answers. Know-it-alls ultimately want to be heard, and if you channel their energy the right way, they could become extremely valuable assets to the team. Say things like, "You bring up a great point," or "I never thought of it that way." If you are not opposed to letting them get the recognition they desire, you can win big in the end.

Fig. 4–11. The know-it-all is like a 200-pound encyclopedia; it contains useful information but takes up too much space.

The bottle rocket. Bottle rockets like to think of themselves as know-it-alls, but they do not have nearly enough knowledge to actually back it up. They are called *bottle rockets* because bottle rockets promise a spectacular firework display but ultimately disappoint. They do not deliver the goods. Although they speak with confidence, they often spew out inaccuracies that could ultimately lead your team in the wrong direction.

How to cope with the bottle rocket:

- Bottle rockets need to feel valuable, but if you give them too much praise, they may start to believe their own press.

- State the facts, but do it in a way that allows them to save face.

How to motivate the bottle rocket: As with the know-it-all, sometimes a little assurance is just what the doctor ordered. Telling bottle rockets that you liked the way they handled a specific task is music to their ears. If you team up these type of people with one of your key players, they tend to stay on track with the assigned task.

The avoider. Avoiders are the stereotypical wallflowers. They are unresponsive people who sit quietly in the corner and prefer to work alone. When forced to work on a team, they speak in superficial terms and tend to do little more than validate what someone else has already said. They do this as a defense mechanism to protect themselves from saying something that others may consider foolish.

How to cope with the avoider:

- Encourage them to get involved with the project early on and try to keep them involved.
- Do not threaten them. They are motivated by the need for security. Fear debilitates them.

How to motivate the avoider: When constructive criticism is necessary, begin by assuring them that their jobs are not at risk. As with the compulsive pleaser, for best results, you should utilize the sandwich technique. This technique is effective with all personality types, but especially effective with avoiders and compulsive pleasers.

The whiner. The world is not perfect, but whiners believe it should be. They find fault with everything and everyone. Even when they are put in a perfect situation, they will find ways to make it imperfect. In their eyes, no one measures up, and they will complain to you because they expect you to solve these problems for them.

How to cope with the whiner:

- Listen to their points of view, but do not agree with them or apologize if they are wrong.
- Switch their energy to problem solving.
- Get them involved with the process; if they feel they have ownership in a project, they tend not to complain.

How to motivate the whiner: Tell them you are okay if they bring you a problem as long as they also bring you a solution. When you channel their energy from the problem to the solution, they may surprise you with some creative problem solving. You will also find it helpful to begin your day by looking them straight in the eyes, smiling, and enthusiastically saying, "I can tell already that we're going to have a great and productive day today! Can't you?"

The negativist. Negativists are pessimistic about everything. They rarely say anything positive, yet they are quick to point out why something will not work. They are generally inflexible, and even though they never admit that anything is going the right way, they always resist change.

How to cope with the negativist:

- Avoid getting sucked into their world.
- Be clear with your own realistic expectations and optimism.

How to motivate the negativist: Inevitably, there will come a time in a firefighter's career where he or she will respond to a fire that is too large for the resources available. The only option at that point is to isolate the fire and protect the exposures. Negativists are like a fire that is out of control. It is worth taking time to discover whether or not they are dealing with personal issues or just their individual personalities. If it is more than a passing phase, consider isolating or replacing negativists with others who bring positive energy to the team.

> *Diplomacy is the art of letting someone have your way.*
>
> —Daniele Varè

Unaware, unable, or unwilling

Before you place a team member in the problem personality category, make sure you take time to determine if he or she belongs there. Consider it your job as a leader to determine whether the individual is unaware of the problematic behavior, is unable to change the behavior, or is unwilling to change.

I call this the *3U method* and cover this technique thoroughly in *Step Up and Lead*. This philosophy has helped me be a more effective fire service leader, and I highly recommend it to anyone leading any type of team. As you gather the facts, you can begin contemplating your course of action. You can do this by thinking: "If they're *unaware*, I will . . . "; "If they're *unable*, I will . . . "; or "If they're *unwilling*, I will . . . " and then act accordingly.

People who are unaware that their behavior has changed may be dealing with personal issues, and a simple talk may be exactly what the doctor ordered. When team members are unable to change problematic ways, you may have to provide some form of training or professional counseling. Either way, these two categories are much easier to deal with than the third. When people are unwilling to change, you are dealing with a much more serious situation. People on your team who are unwilling are insubordinate, and they must be dealt with properly before their attitudes infect others.

The meeting of two personalities is like the contact of two chemical substances; if there is any reaction, both are transformed.

—*Carl Gustav Jung*

Accountability

It was 2:45 AM on a cold December morning, and we were on the heels of the biggest snowstorm in years. Unfortunately, my prediction from the night before came true.

Step Up Your Teamwork

"Something big is going to happen today," I said to the firefighter I was commuting with. On our way to work, we slowly inched our way past vacant, stranded cars that were half buried in snow. It was 20°F, with almost 30 inches of snow on the ground from accumulated snowfalls. Only the main streets in town had been plowed, making it impossible to access most of the 100 or more side streets in town. Furthermore, most of the hydrants in town were fully covered by snow, making water supply a major concern. Crew members from the preceding shift had worked hard to expose the hydrants on the corners of the streets that had been plowed. Our only option was to hope for the best until sunrise, when the town would have time to plow the side streets and we would have a chance to clear the remaining hydrants. That was the plan, but given the circumstances, it seemed only appropriate that Murphy's Law was about to come into play once again.

Our dispatcher sent in the alarm tones, followed by the announcement, "Full assignment District 1. Report of a working fire."

As I climbed into my car, my commuting pal looked at me and said, "You called it, Chief."

The structure was not a single-family dwelling; it was a huge nursing home with more than 100 elderly residents. An attached structure housed a rehabilitation center. Engine 1 arrived on scene prior to me and confirmed a moderate smoke condition inside the structure on the second and third floors. I immediately called for a second alarm, knowing that the life hazards and water supply challenges were enough to justify the additional personnel and resources even if the fire turned out to be easily containable.

Luckily, the street this structure was on is one of the main streets in our community and was one of the few streets that had been plowed earlier. The closest hydrant was cleared for access, but this large facility covered the full length of the block, and the main emergency access point for elderly residents was on one of the side streets. That street was only partially cleared of snow, making it accessible only by large, heavy vehicles, such as fire engines, or by other vehicles with four-wheel-drive capabilities.

We were lucky that the fire was confined to a small medical storage closet and was contained by an activated sprinkler head. However, that did not change the fact that we had two floors, the second and third, that were charged with smoke and occupied by more than 100 seniors. Many of the residents could not stand and walk unaided, and more than a few were receiving medications and nutrition intravenously. The supplies that were burning gave off an unfamiliar odor. Since the fire was in a medical storage closet, we had to treat the situation as a hazardous materials incident. We also had to remove the elderly residents because of respiratory concerns.

We set up our command post in the lobby and began to establish our game plan. I was surprised to see all the patients and staff of the attached rehabilitation center, about 40 of them, walk into the lobby area. I was informed that the smoke alarms on their side of the building had activated as well. It was too cold for anyone to remain out in the street, so we assigned one area for them and another for the elderly residents who would have to be removed from the smoke-charged floors above. The total number of people in both buildings exceeded 200. The lobby could hold around 50, but not comfortably. The lobby area quickly filled with firefighters, nurses, and patients, and it did not take long before a large number of additional emergency response personnel from various agencies to begin pouring through the lobby door.

This is where an organized accountability system becomes essential (fig. 4–12). A good accountability system, such as a magnetic command board, allows an IC to track all personnel and agencies working on the scene (fig. 4–13). At this incident, I had to account for personnel on 10 fire apparatus from four departments, 12 ambulances from seven towns, a hazardous materials team, various law enforcement and emergency management personnel, public works employees, a health inspector, a town building inspector, and facility building maintenance and management personnel. This was all in addition to more than 100 elderly residents, 40 rehabilitation occupants, and 12 or more facility staff members. I had not met most of the emergency responders prior to that day. In fact, I was not even working on my shift. I was covering for another chief, so I did not even know the strengths and weaknesses of most of the personnel on duty. There was no time to shake hands and get to know people. We had to get to work quickly to ensure that no lives would be lost.

The first thing I did was set up an accountability board at the entrance of the lobby. I then assigned one firefighter as the incident accountability officer. His job was to help me track personnel and agencies as they entered or exited the facility. No one passed the command post without having interaction with me. When I did interact with them, my goal was to give each member an assignment. Their agency, designation, name, and assignment was then documented on the accountability board so I knew what each person and agency was doing at any given moment.

Fig. 4–12. Large structures such as this former hospital require an organized accountability system to help track personnel and residents.

Police department personnel were assigned the job of assisting with traffic and pedestrian control. One sergeant took on the role of ordering any resources we needed, such as transportation buses and a town vehicle with a plow. I assigned the head of our EMS agency the task of organizing the removal of victims and arranging their transportation, which was being provided by a dozen ambulance companies. One of the facility nurses was assigned as the victim tracking coordinator. Her job was to assure that every patient was accounted for and receiving whatever special medical attention was required. A member of our town health department assisted with any and all health concerns. Building maintenance personnel worked along with our firefighters to assure all building components were controlled (elevator, sprinkler system,

HVAC, alarm system, etc.). A certified electrical contractor and our town building inspector ensured the building was safe from electrical fires that could have resulted from the water that was leaking from the second floor ceiling. The building manager was called to the scene to organize a cleaning contractor to assist with water removal and building repairs that resulted from the fire. The facility administrator assisted with the relocation of more than 50 elderly residents by arranging their transfer to other nearby facilities. As if that was not enough to account for, I still had 10 fire apparatus and a hazardous materials response team, with each assigned different tasks. In other words, without an effective accountability system, there would be nothing but chaos.

Fig. 4–13. An accountability board, when used properly, is a great tool to help keep track of personnel from various agencies.

The point is that you must learn how to delegate assignments and account for your team members if you intend to operate at a high level and prevent team collapse.

How to delegate assignments

When issuing assignments and delegating tasks to team members, in your mind you should answer the basic questions concerning who, what, where, when, why, and how:

- Who am I communicating with, and what is that person's (or team's) skill set?
- What specific task do they need to achieve?
- Where should they go to begin their goal of accomplishing this task?
- When does the task need to be completed?
- Why is this specific task essential for us to achieve our ultimate goals?
- How should they go about completing this task?

Once those questions are answered and the assignments are given, be sure that every team member understands the mission. Next, document who was given what assignment in a place where it can be referenced quickly and easily; for instance, I use an erasable white board (fig. 4–14). This will come in handy as a reminder of what tasks are being worked on. Anytime you have to make contact with a team member, all you have to do is reference your accountability board. Require periodic progress reports from each person or team so you can track their progress.

Whenever you assign a task, make sure you provide the person with the tools and resources needed to complete the task. Don't give out an assignment without giving your team what they need to do the job. If you make a habit of assigning people tasks without adequate resources, they will think you are either a poor leader or setting them up for failure. This is an example of how accountability on the fireground should work, and the story above illustrates the system we used to save lives at the nursing home fire. Fortunately, the fire was contained early. Even so, the building was charged with smoke and some residents had compromised respiratory systems. Combined with the fact that it was below freezing outside, there could easily have been casualties at this incident.

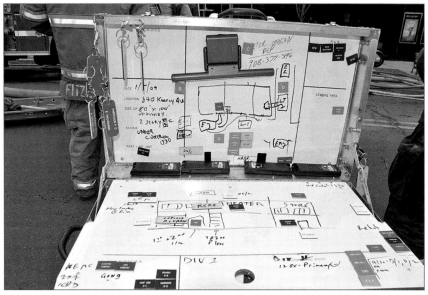

Fig. 4–14. Use some form of an accountability board to document the assignments you give people. This act should be done on the fireground as well as in the soft environment. Doing so will help you stay organized and focused.

Individual accountability

According to BusinessDictionary.com, *accountability* is defined as "the obligation of an individual or organization to account for its activities, accept responsibility for them, and to disclose the results in a transparent manner." Accountability on an individual level is the willingness to accept responsibility or to account for one's personal actions.

In the simplest terms, accountability results from having to answer to someone. I have never met a truly successful person in any area of life who was not accountable to others. This includes business owners, investors, athletes, spouses, and firefighters. Being accountable is important because it makes you work harder, not just for yourself, but for the good of the overall team.

There are three main areas in which fire service leaders are concerned with personal accountability. The first is on the fireground, which was covered in the previous story, and the second is in the fire station. The third is an area in which many second-rate companies would not concern themselves—their actions while off duty.

Accountability is not an unfamiliar term to firefighters. Sadly, more than 100 men and women who take the oath lose their lives in the line of duty each year. Some of those firefighters make the ultimate sacrifice because they became disoriented or trapped inside a burning structure. If an accountability system is in place and the other crews working on scene know where a firefighter is when a Mayday is called, the distressed firefighter has a chance. If not, the chances of making a rescue diminish greatly. The bottom line is that accountability in our profession can potentially be the difference between life and death.

On an individual level, when a person is accountable to another person, or to a measurable system of some sort, that person has a greater chance of achieving a higher level of success. This is mainly because that person knows that if the task is not accomplished as promised, he or she will eventually have to explain why.

When a person is accountable to another person, or to a measurable system of some sort, that person has a greater chance of achieving a higher level of success.

Soft environment accountability

In the fire station and during times when we are not answering the call, great team leaders still understand the importance of accountability. When we are working in a soft (or nonhostile) work environment, we still have rules that should be enforced on a daily basis. It is necessary to have rules and regulations for the self-directed freelancing members who think they can do what they want whenever they want. We have to ensure that specific and important rules are uniformly applied and enforced. Some people scoff at the idea of enforcing rules. Perhaps they should be reminded that they signed an agreement to obey our rules and regulations. The first breakdown in a fire department is the degradation of standards.

One of the things I like about rules is that they help me gauge a person's work ethic, and in some cases, his or her integrity. The way people adhere to rules will tell you a lot about them and how they respond to tasks and challenges. If I observe people consistently finding ways to avoid what they are supposed to be doing, I know they are unreliable. It is not just a matter of my opinion; their actions will lead others to that conclusion also.

Clearly established rules will help you institute a higher level of accountability. In the absence of rules, people make their own. Many companies have failed because good people were inadvertently working against each other. Others have failed because the rules were not clearly established or followed. As a team leader, it will be your responsibility to address the problem when rules are broken and people begin to freelance.

In the absence of rules, people make their own.

Off-duty accountability

Firefighters need to be accountable for their actions, whether they are on or off duty. A firefighter belongs to one of the largest and most respected clubs in the world. Once firefighters take the oath, they become representatives of the fire service. If they get in trouble or act in ways that embarrass themselves or their organization, they run the risk of giving our entire profession a black eye.

I am not just talking about the individual who wears his fire T-shirt to a local tavern and has a confrontation with another patron, although that is never good. I'm talking about *any time* a member of the fire service does something illegal, immoral, or unethical. This includes the way a person acts on social media sites. Yes, we have first amendment rights just like anyone else, but many public service workers have been suspended, demoted, or fired for posting comments that reflected poorly on their organization. Do not make this foolish mistake. Instead, be a model for integrity and teach others on your team to be the same. It may not be a popular thought, but remember that freedom of speech is not always freedom from consequences. It is not easy to do the right thing all the time. It is hard work, but you have an obligation to set the standard and example for others to follow. The fire service has spent many years developing an image of trust, respect, and loyalty. We cannot afford to ruin this by allowing members of our team to tarnish that reputation. If you intend to develop a successful team, make sure they understand that their actions reflect on the overall image of your organization. Set the right example; not just for now, but for future generations.

Freedom of speech is not always freedom from consequences.

Accountability partners

Few professions have the uniqueness that rivals that of the fire service in terms of accountability. You have just read about fire scene, individual, soft environment, and off-duty accountability. Now let's discuss the importance of accountability partners.

Imagine you are the firefighter in this scenario. Ten minutes before midnight, your crew is dispatched to an activated fire alarm at 118 Main Street. You get out of bed, throw on your clothes and bunker gear, and jump on the ladder truck, along with two other firefighters. The siren echoes throughout the empty streets. The dispatcher comes over the radio and reports multiple calls, confirming it's a fire. You knew it was, because you could smell it from six blocks away.

Chapter 4 — Preventing Team Collapse

The ladder turns down the street, and you see a massive glow at the end of the block. You know it's going to be a tough one. The butterflies start, but you quickly push the negative thoughts out of your head and begin to psych yourself up. For a brief moment, you think about your spouse and children, who are home sleeping. You imagine their peaceful faces and take a moment, just a moment, to ask God to protect them.

The engine company pulls a hoseline from the apparatus and begins advancing it inside the structure. People are in the street, waving and pointing toward the house. All you hear are sirens from other approaching apparatus, but you can tell by the way they are acting that someone is trapped inside.

The incident commander turns from the people he is talking with, walks up to you and your partner, and says, "Follow Engine 3 up to the second floor. An 11-year-old child is unaccounted for. His bedroom is toward the front of the building."

Your child is 11.

You don and tighten your SCBA mask and follow the line up to the second floor, along with your partner. The driver stays with the apparatus in order to raise the aerial to the roof. At the top of the stairs, the line is being stretched toward the back of the structure, where the fire is. You quickly communicate with the engine captain and nozzle man and let them know you are heading in the opposite direction to search for the boy. You and your partner start crawling down the dark hall. Visibility is zero. You can hear the faint sound of crackling fire, and approaching sirens. Then suddenly, it all fades and the only thing you hear is your own breath. You continue moving down the hall until you reach the bedroom door.

Before entering, you lean toward your partner and yell through your mask, "I'll go left, you go right!"

Through the smoke, you can barely make eye contact, even though your masks are practically touching. It occurs to you that all the training you have been through comes down to this moment, and although you belong to a team, the only person you can rely on is the other firefighter in the room with you—your partner.

A situation like this happens every single day in America. You can have the most highly functioning team in the world, but when you find yourself crawling through a burning structure with one other firefighter, you become accountable to each other in a way few people will ever understand (fig. 4–15).

Fig. 4–15. Few understand the importance of having a good accountability partner the way firefighters about to enter a structure fire do. *Courtesy:* Cindy Rashkin.

An accountability partner will encourage you, push you, help you, and commit to you. If we are going to commit to each other on such a high level at a fire, we should be willing to keep that commitment after the incident is over. Accountability partners can help each other stay on track toward reaching their goals and creating the desired organizational culture.

You need to have an accountability partner to help you stay focused. Once you find the right person, it is time to stop procrastinating, stop rationalizing, and stop making excuses. If you are serious about changing and improving, the best way to do it is to take action. When you are only accountable to yourself, you tend to make excuses, but when you become accountable to someone else, you are more likely to follow through on your promises. It is much more difficult to make excuses and rationalize if you have an accountability partner who refuses to let you get away with it.

Interlocking accountability

If you truly want to enhance your teamwork, you will want your entire team to become accountable to each other. This is called *interlocking accountability*. Here is how it works. First, everyone must agree to hold each other accountable. Whenever someone does something right, the others acknowledge it. Whenever something goes wrong, or someone drops the ball, instead of blaming that person or judging, acknowledge the problem so it can be corrected. Do this without assigning blame. Support the person so positive change and progress can be made. Do not become the type of organization that singles out a person and uses that individual as a scapegoat. I have seen teams that relentless pick on one person. It reminds me of a bunch of bullies picking on the weak kid. Be better than that. Put the team first. We have to, because when you get right down to it, our lives depend on it.

Put the team first. We have to, because when you get right down to it, our lives depend on it.

Climate

The climate in which your team works has a lot to do with their success, or lack thereof. When our department was in the process of hiring our first female firefighter, new policies had to be written. Up until that point, everything from the bathrooms and sleeping quarters to the type of "material" some members had hanging in their lockers was male-centered.

After drafting these new rules and discussing behavior, we embarked on a much-needed climate shift. Now, everything each member said, did, watched, and read had to align with the new culture of our organization. There was greater emphasis placed on equality, respect, manners, and in the end, we were all better for it. The main lesson our members learned during this time was simple—what's bad for women is bad for our team.

Dialogue, Discussion, and Debate

Open lines of communication are essential. When people feel like their voices are being heard, they believe they are valued members of the team. When a problem occurs, and a team meeting is in order, sometimes all that is needed to fix the problem is dialog, discussion, and debate. It is true that there are times when you will be beyond this point and at a crossroads where immediate action needs to be taken without discussion. However, most people forget the importance of involving others in the decision-making process. When people feel valued, they begin to think and act as if they have ownership in the team.

There is a plague in today's society called *entitlement mentality*, and it is embodied by people who want all the rewards without having to do the required work. When your team is comprised of people like this, failure is inevitable. One way to prevent this disease from spreading to your team is to create the opposite of entitlement mentality—ownership mentality.

Ownership mentality can be deliberately created by encouraging people to get involved in making decisions that are best for the overall mission of your team. Even when their suggestions are not acted upon because you feel there is a better option, they will feel more appreciated, and that is often the exact thing the team members need.

There is a plague in today's society called entitlement mentality.

Dumb Things Firefighters Say

You have heard them all. Maybe you have even said some of these things. I am talking about fire service staples like the following:

"Don't worry, this is just a routine fire."

"All you have to do is put the wet stuff on the red stuff."

"Stay in until you can't stay in anymore."

"Because that's the way we've always done it."

"Keep your mouth shut, Probie."

"Books don't put fires out. Why are you wasting your time reading that?"

I was speaking to a group of sales professionals about leadership when one sales manager put his head in his hands and said, "I just realized what a lousy trainer I've been." His comment surprised me because he was the top trainer in the room. He went on to say, "Every time a new sales rep asked me what he should do, I replied, 'Make your numbers.' That was the dumbest thing I could have said."

Most of the others in the room were confused. His answer to the new sales rep's question was not an unusual one; they had all said it before. However, the man was right. New salespeople are well aware that they have a quota, and their compensation is tied to that quota. They do not need someone to state the obvious. "Making your number" has likely been the only thing on their minds since the moment they took the job. What they were asking was, "What actions should I take in order to reach my quota?"

Now imagine the new firefighter who is sitting at the table with a bunch of experienced firefighters, you being one of them, and the advice people are giving is, "All you have to do is put the wet stuff on the red stuff." The young firefighter is well aware of the fact that water puts fire out, but giving such poor advice will do nothing other than weaken your team. You have a clean slate to work with. Do not dumb down this person with overused, and often incorrect, one-liners. Spend time with them. Educate them. Train them. Make sure they learn the right way and the wrong way. The person you give advice to today may be the one you have to rely on to save your life tomorrow.

The person you give advice to today may be the one you have to rely on to save your life tomorrow.

Multiple Alarms—Call for Help Early

On the fireground, when the incident becomes too large and difficult to fight with the resources we have on scene, we have two choices: stand there and watch it burn or call for multiple alarms. Obviously, letting it burn is not an option (don't confuse *letting it burn* with a *defensive strategy*). Our only real option is the second. There are two main reasons why we call for multiple alarms:

1. To obtain more personnel, equipment, and resources to aid in physical firefighting operations

2. To obtain more personnel to fill command staff positions

As your team takes on more ambitious goals, you may find yourself in a situation where you are overwhelmed and need additional resources. Do not hesitate to recognize that you are going to need assistance and call for help early. A smart team member will gain respect from others by anticipating the need for extra help and addressing that need before it is too late. This is a critical concept that can help you prevent team collapse.

The National Institute of Occupational Safety and Health (NIOSH) 5 is a list of the top five causes of line-of-duty deaths (LODD).[1] Most firefighters would immediately think of things like heart attacks, roadway accidents, or asphyxiation. Those, however, are actually the results of the causative factors. A re-examination of the NIOSH LODD reports shows that the top five operational causes of LODD all surround command. They are:

1. Improper risk assessment (poor size-up)
2. Lack of command
3. Lack of accountability
4. Inadequate communications
5. Lack of SOPs (failure to follow SOPs)

This list is an eye opener. It should be all the proof you need to validate the concept that team leaders cannot afford to be reactive; they must be proactive.

When you find yourself in over your head, call for help. Don't try to solve every problem by yourself. You are only one person, with one mind. When you have competent people around you, get them involved in helping you make critical decisions (fig. 4–16).

Fig. 4–16. Call for additional resources when you feel like the situation is becoming larger than you or your team can control. Do not wait until it is too late.
Courtesy: Ron Jeffers.

Calling for help is something we should do on and off the fireground when we need assistance. If you are unsure what actions to take when confronted with a unique situation, use a lifeline and phone a friend. That person may bring a fresh perspective that will enable you to see things from another angle. At the very least, a friend will help you work out some of the challenges you are dealing with so you can come up with the best possible solutions, prevent team collapse, and attain true team success.

Reference

1. Anthony Kastros, "Mastering Fireground Command: Calming the Chaos," *Fire Engineering* (Tulsa, OK: PennWell, March 1, 2011).

BUILDING YOUR LEGACY

5

Have you ever wondered what your legacy will be? Every individual and every team leaves a legacy. Your legacy will be remembered long after you are gone. What do you want to be known for? Being a key player on a winning team has always been one of the things at the top of my list. I love being on championship-caliber teams, and I am thankful that I have had the opportunity to be part of several great teams in my lifetime. From a nationally ranked, hall-of-fame high school rowing team, to a highly productive top-producing sales team, to a fulfilling career in the greatest team profession known on earth—firefighting—I have been blessed to have experienced team success.

Ironically, early on I doubted I could achieve success in any of these areas. One of the biggest mistakes I made as a younger man was to compare my abilities to others around me who were successful, and my abilities often did not stack up to theirs. However, it made me work harder. One of my favorite sayings is, "No matter how many mistakes you make or how slowly you progress, you are still way ahead of everyone who isn't trying." That quote has carried me through many bouts of self-doubt and has helped me overcome a terrible inferiority complex. If you are anything like me, there is hope.

I do not know what you want your team or personal legacy to be. It is my sincere hope that this book, and especially this section, will help you answer the question, "What do you want to be known for?" Since we have come this far in our journey together, stay with me for one more chapter and give me the opportunity to share some thoughts and philosophies that might help you achieve success in the areas that are most important to you.

No matter how many mistakes you make or how slowly you progress, you are still way ahead of everyone who isn't trying.

—*Unknown*

Turn Adversity into Advantage

It is inevitable. Once your team is established, your mission is clear, and you set your goals and objectives, problems will surface. It would be easy to poll your team members to see how everyone thinks a problem should be addressed, in an effort to lead by committee, but doing so will often lead to a stalemate. Synergy and momentum will be lost, and feelings will be hurt. So, what is the answer? What do you do? Certainly, you want your team involved in the decision-making process, but it is time to step up. When hard decisions must be made, you have to be willing to make them.

We all have problems. The difference between successful teams and unsuccessful teams is how you look at them. Do problems and challenges derail your team? Or are your members so focused on team goals that they consider problems merely obstacles to overcome? The level of success a team achieves is in direct proportion to their ability to overcome those obstacles. Show me a championship team in any arena, and I will show you a team that has mastered the art of overcoming.

The 2004 American League Championship Series was played between the Boston Red Sox and New York Yankees. During that showdown, the Red Sox became the first team in MLB history to lose the first three games and win the next four to win a seven-game series. I watched one team (the Yankees) collapse, as their bitter rivals, the Red Sox, came back against incredible odds to not only win the AL championship, but also the World Series. Watching the Red Sox overcome adversity to beat the Yankees in that series was nothing short of incredible—painful for Yankee fans like me, but incredible.

What adversity is your team facing? Are you 0–3 in your professional version of a championship series? Have work-related challenges been beating up you and your team lately? Have you recently received complaints from valued customers? Are you struggling to keep up with professional demands? Whatever the problem is, put it in perspective. Remember that problem solving is why your team exists (fig. 5–1).

Fig. 5–1. Your team exists to solve problems. *Courtesy:* Ron Jeffers.

In his book *Obstacles Welcome: Turn Adversity to Advantage in Business and Life*, AT&T Mobility President and CEO Ralph de la Vega lists the following as his four pillars of success:

1. You must plan for success. Hope is not a strategy.
2. You must take calculated risks to get into position to achieve big goals.
3. You must learn to recognize opportunities. The key is to realize that most important opportunities are problems waiting to be solved.
4. Embrace and overcome obstacles. They make you stronger, wiser, and more capable.

Adversity will test your team, but it is important to find ways to overcome the challenges you will encounter. Although adversity has the ability to destroy your team, it also presents you with the opportunity to get stronger. There are many reasons why it is important to cope well with setbacks and overcome adversity. Among the more important reasons are the following:

1. Prevents discouragement
2. Prevents a decrease in morale
3. Shows that the team can stay focused and continue to move forward
4. Enables your organization to maintain credibility within your community
5. Helps the team develop a reference base for how to overcome future adversity
6. Further develops the team's ability to work under pressure

In addition, overcoming adversity will enhance your leadership ability and increase the team's faith in you. The real measure of a leader becomes apparent when things are not going well.

Every adversity, every failure, every heartache carries with it the seed of an equal or greater benefit.

—Napoleon Hill

Failure Isn't Fatal

A coworker of mine was driving through his hometown. He was off duty but smelled the familiar odor of a house fire. He looked down a few side streets and eventually saw a column of smoke. He drove to the burning structure and quickly realized the fire department had not been called yet.

A woman was in front of the house frantically waving her arms to capture his attention. He pulled over and jumped out of his car. It was a small two-family home. The fire was on the first floor.

"Is anybody in there?" he asked.

"Yes, I think so. He lives alone," the woman answered.

"Call 9-1-1!" he directed, before kicking in the first floor door and entering the structure.

He quickly dropped to the floor and started crawling. The smoke and heat were noticeable, but he kept moving deeper into the apartment until he came upon an unconscious man. The flames were rolling above his head, and the fire was growing. Conditions were changing faster than he anticipated, and the heat was becoming unbearable. He briefly considered turning around and bailing out, but he did not. Although it took every bit of strength and guts he could muster, he was able to drag the man out to safety just as the fire department pulled onto the scene. Alone, and without gear or the protection of a hoseline, his actions were very risky and ill-advised and could have cost him his life. So why did he do it?

At a working fire two years earlier, he had failed to find a woman who was in the room where he conducted a primary search. If you listened to him share the details of that event, you would know it was not his fault. However, for two years he had carried a heavy burden of feeling like he had failed. When he was finally presented with another chance, he was determined to succeed. My friend used what he perceived as a personal defeat to propel himself to do something he did not know he was capable of doing.

> *If you learn from defeat, you haven't really lost.*
>
> —*Zig Ziglar*

How do you perceive failure? I want to encourage you to look at failure for what it is—an opportunity to grow and change. Teach your team how to put failure into perspective. Next time you are in a room full of your peers, look around and pick out the person whose team has achieved the highest level of success. There is a great possibility that you will also be looking at the person in that room who has failed the most.

You have two choices when you experience failure: give up or get up.

Thomas Watson, Sr. was the chairman and CEO of International Business Machines (IBM) from 1914 to 1956 as the business grew into an international phenomenon. Watson developed IBM's distinctive management style and corporate culture, and turned the company into a highly effective sales organization. A leading self-made industrialist, he was one of the richest men of his time and was called the world's greatest salesman when he died in 1956. Watson was an incredibly accomplished man. He was also quoted as saying, "The key to success is massive failure."

Look through the history books at some of the most successful political, business, and sports icons in our history: Abraham Lincoln, Sam Walton, Col. Sanders, Thomas Edison, and Michael Jordon. What one thing did they all have in common? They each had failed repeatedly.

Lucille Ball was another failure. Before starring in *I Love Lucy*, Ball was widely regarded as a failed actress and a B movie star. Even her drama instructors did not feel she could make it and told her to try another profession. She, of course, proved them all wrong. By the end of her career, Ball had 13 Emmy nominations and four wins, also earning the Lifetime Achievement Award from the Kennedy Center Honors.

Based on these examples, it is easy to conclude that the only way to achieve more success is to experience more failure. Once you understand this, you should no longer be afraid of failure. Instead, accept it for what it is—a necessary part of the success process.

Failure is a learning tool. When it occurs on your team, you can ask questions like the following:

- What brought about the failure?
- Did my actions play a part in this outcome?
- What do we need to do to turn failure into success?
- What can I do to ensure I will make the right decisions next time?

Failure builds character. You can choose to throw in the towel and crawl into a hole, or you can learn from the experience, gain confidence, press forward, train on your weaknesses, and become the person that you ideally want to be. Failure isn't fatal if you continue to learn from it and train on your weaknesses (fig. 5–2).

Failure comes with benefits. For example, it encourages lateral thinking, forces you to be honest with yourself, gives you valuable experience, reveals your weaknesses, and keeps you humble. Admit your own mistakes and encourage others to do the same. We all make mistakes from the day we are born until the day we die. Babies learn to walk by falling down. Even an MLB Hall-of-Fame hitter misses the ball 7 out of 10 times. Every experienced fire chief I know has lost a structure at least once. You will make mistakes. Just make sure you learn from them.

We all want to be successful, but teams that experience too much success early in the game often develop a false confidence. This occurs quite often in business and sports. When people think they are better than they are, they tend to slack off. This can have disastrous consequences in a profession like ours where concentration and situational awareness are so vitally important.

Step Up Your Teamwork

Fig. 5–2. You must learn from your mistakes and train on your weaknesses.

Learn from the past, but don't live in it.
Don't let one bad decision wipe out all the good
you have accomplished.

—*Frank Viscuso*

Courage under Fire

Some people may foolishly think firefighters are fearless, but this is a huge misconception. Even though firefighters are known for their courage, the majority of men and women who do the job do not know if they are truly courageous until the day comes when they are put to the test. Even though a person may have consciously chosen a career in the fire service, how could anyone know if the raw determination or audacity to do the job is there until the moment comes where the person is expected to put it all on the line?

What we discover along our journey is that firefighters are not without fear. Many have simply found a way to master it. We are a group of individuals who consistently find ourselves in situations that force us to push ourselves. We go from a relaxing lunch to a battlefield within seconds; from a group of individuals hanging out to a high-performance team in an instant (fig. 5–3). To illustrate the point, consider that a group of people stuck in an elevator that has stopped moving is just a group of people stuck in an elevator. However, a group of people stuck in that same elevator when the building is on fire quickly becomes a team!

Fig. 5–3. Firefighters often go from a relaxing lunch to a battlefield within seconds. *Courtesy:* Ron Jeffers.

We are a team that is expected to operate under the most hostile conditions imaginable. Danger is always present, but as Jean Paul Friedrich Richter said, "Courage consists not in blindly overlooking danger, but in seeing it, and conquering it." Even so, do not be fooled into thinking that firefighters are more courageous than other people. The salesperson who is afraid of rejection but spends the day making cold calls is courageous. The person who is afraid to speak in public but earns a living making presentations is courageous. Parents who decide to home school their children are courageous. It does not matter what a person does for a profession, the fact is, there will come a time in everyone's life when he or she will have to display a higher level of courage that seemed impossible. Do not feel like you are incapable of succeeding because you are afraid. We all experience fear. The key is to feel the fear and do the job at hand anyway.

You may be facing a major challenge in your personal life right now. Your back is up against the wall, and you feel tired and worn out. Maybe you are on the verge of giving up. Don't quit! It is at times like these when you discover how strong you really are. Like any other human being, you have the ability to put fear aside and muster enough strength to get the job done. You can be courageous under fire.

How do you develop courage under fire? Having a purpose is the best way to overcome fear. Knowing what you must accomplish and why you must go through discomfort to achieve your goal is paramount. Without vision and determination, most people quit at the first sign of resistance. I personally draw courage from my family. I think about how important it is to make my parents proud of me. My wife and children are, without a doubt, a major source of strength for me. Most parents would agree they would do anything for their children. This includes enduring discomfort to ensure they can live a more fulfilling life. That does it for me; that is where I draw my inner strength to push harder. The question I want to ask you is, "Where do you draw your strength?"

Take a few minutes to consider your answer. Once you determine what is important to you, it will be easier for you to push fear aside and press forward in challenging times. History is full of many great examples of courage, from Jesus Christ to George Washington, Rosa Parks, William Wallace, Nelson Mandela, Helen Keller, Mother Teresa, and Mahatma Gandhi. Many historically notable people have left a legacy of courage, wisdom, and endurance. Anyone can give up; it is the easiest thing in the world to do. But to hold it together when everyone else would understand if you fell apart, that is true strength. That is courage, and people want to follow courageous leaders.

He who is not courageous enough to take risks will accomplish nothing in life.

—Muhammad Ali

Combustible Courage

When one or more members of a team display a noticeable level of courage, their actions tend to empower others around them. Just like a person's attitude, courage can also be contagious.

Combustible courage is a term I use to describe the phenomenon of how one person's actions can elevate the audacity of an entire team. A Spanish conquistador named Hernán Cortés understood this very well. Legend has it that in 1519, Cortés assembled an army of more than 500 and set out to take the world's richest treasure, held by the Aztec empire for 600 years. Army after army had failed to take this treasure. Cortés knew what he was up against and could sense the doubt creeping into the minds of the men as their ships sailed toward their destination. One of the things Cortés did to keep their minds focused on the prize was speak to his soldiers about how their lives were going to improve after they acquired the treasure; their families would never have to struggle again. However, this only worked for a short time because their food supply was running low. He knew that when they landed on Veracruz, they would be tired and hungry. He knew the natural instincts of his army, likely the same as any other group of men faced by such overwhelming odds. He understood that once they landed and faced the Aztecs, his troops would realize for the first time that they were severely outnumbered and outmatched. Cortés believed his army would try to retreat to the boats. He needed to make their level of commitment greater than that of the defeated armies that had preceded them. He needed that one extra degree.

When he reached his destination, Cortés gathered his army on the day of the battle. I am sure they expected a strategic talk or perhaps a motivational speech, but he knew that would not be enough. Instead he turned to men with torches and said, "Burn the boats." The message he sent to his team was obvious. If they were going back home to their families, it would on the enemies' boats.

Cortés and his army defeated and conquered the Aztecs and took the treasure, something no other army before had been able to do. His level of courage and commitment provided the example the others needed in order to attain their one extra degree, the difference between hot water and steam.

Out of every one hundred men, ten shouldn't even be there, eighty are just targets, nine are the real fighters, and we are lucky to have them, for they make the battle. Ah, but the one, one is a warrior, and he will bring the others back.

—*Heraclitus*

What's Your One Degree of Difference?

Let me share a quick story with you about how I discovered my one extra degree. After high school I received a rowing scholarship to attend the University of Charleston, West Virginia (fig. 5–4). Those of you who are familiar with the sport of crew understand why many people consider it to be the ultimate team sport. When eight people are rowing as one, every stroke of the oars must be synchronized to the count of the coxswain. Everyone must exert equal effort. To reach the finish line, everyone must row in unison for the duration of the course, which is typically a mile or longer. To win the race, they must maintain discipline, stamina, and focus from the first stoke to the last. If one person slacks or falls out of sync with the others, their team will lose.

Fig. 5–4. The University of Charleston on the Kanawha River in West Virginia, where I discovered my one degree of difference.

Crew teams are divided into several categories and divisions. The most common are singles, doubles, foursomes, and eights, both with and without a coxswain, who steers the shell and encourages rowers. I had been offered a partial scholarship to attend Charleston and row as a lightweight, which was a crew that maintained an average weight of less than 160 pounds. I was an above-average rower, but we had a talented crew, and on this team I was no better than any other rower. Even so, I wanted to make the varsity team as a freshman. The problem was that I was a starboard rower, or one of the rowers with oars on the right side of the boat, and all four starboard positions were filled with returning athletes. At first it bothered me that I would not have a chance to challenge for a spot on the boat, but then it occurred to me that there were only three positions filled on the port side. I could challenge for the fourth position on that side if I wanted. The only problem, and it was a big one, was that I had never rowed port in my life.

Rowing port was like learning how to write with your nondominant hand. It sounded simple, but in reality, switching sides was like learning how to walk again. Still, I wanted to be on that team, so without telling anybody, I put together a game plan and begin implementing it. Every day we met for practice at 5:30 AM, where I spent a couple of hours rowing with the freshman team on the Kanawha River. Then the team would pack up and head to breakfast before our classes began.

Instead of going to breakfast with the rest of the team, I waited for the last person to leave. Then I took out a portable practice device. This was a unit designed to be placed on a dock, and it allowed one person to sit in the seat, put an oar in the water, and practice his technique. An individual could then practice consecutive full strokes, using an oar with holes in it to prevent him from pushing himself off the dock into the river.

Just setting up this unit was a tedious and time-consuming job. It was awkward to carry and weighed more than 40 pounds. The river was much lower than the university boathouse, and the ramp was steep and long. I had to make two trips to carry the unit and the oar down, and another two trips to carry them back when I was done. This was a commitment in and of itself, considering I was already exhausted from practice.

Once the portable unit was in place, I sat down and began to teach myself how to be a port rower. No coach, no teammate cheering me on, and no guarantee that what I was doing was going to get me the position I wanted. I did this ritual every day for two weeks. At first, I could not seem to get the rhythm or technique down. It felt like I had never rowed a day in my life. Leaning my body in the opposite direction and twisting my opposite wrist to make the oar feather challenged my body and mind. After a couple weeks, however, it started to feel natural. I had no problem putting two hours of practice in with my team as a starboard rower, and then switching gears and completing another 1,000 strokes with the portable unit by myself.

I will never forget the day we were rowing and the varsity head coach pulled up next to the freshman and junior varsity boats and said, "Would anybody like to challenge a varsity rower for the remaining port position this season?"

I raised my hand.

"I admire your ambition, Viscuso, but you're a starboard rower," he said, with more than a slight hint of sarcasm.

Everyone laughed. My hand remained up in the air.

"You want to challenge?" he asked again.

"Yes," I replied.

Over the next two days, I competed against two other rowers in a series of tests, from ergometer times, to strength tests, to techniques, and I won the spot. Later in the season, my coach admitted that he knew what I was doing after practice, and it made an impression on him. He was the first one to say to me, "The difference between ordinary and *extra*ordinary is just a little *extra*." In other words, one more degree made the difference.

I share that story with you because that was the experience that taught me about the extra degree. I used that experience to excel in other areas of my life, including my career in the fire service. What is it that you want to accomplish? Are you trying to make the team? Are you serious about advancing your career? Do you want to improve the organization you are leading? Whatever you are trying to accomplish, you may be only one degree away from achieving your dream.

What is *your* one degree of difference?

*The man who rows the boat
doesn't have time to rock it.*

—*Unknown*

It's Not Over until We Win

Motivational speaker Les Brown once gave a powerful talk about perseverance. The climax of his talk came when he said the words, "It's not over until I win!" As soon as I heard those empowering words, I began repeating them every time I encountered my own challenges and obstacles. I drew strength from those words and wanted pass that on to the other members on my sales team. All we had to do is change the word *I* to *we*.

When I began developing this team, I met with a number of friends and shared a simple business plan with them. Some decided to participate, some did not. My intent was never to try to convince people to join the team. I only wanted to share the concept with people I was interested in working with. Although my friend Michael was only 21 years old, he was the exactly the type of person I wanted on the team. Unfortunately, the timing was not right, but he assured me that he would let me know when it was.

A few months later, I met with three engineers who were looking to supplement their income. One was already a member of the team. The other two were impressed with the results the team member was getting from his part-time work, so they wanted to look into it. Normally when people look into something new, they study, listen, and decide if it would work for them. However, when most highly educated engineers look into something, they study, listen, and then analyze whether or not it would work at all. The income some of my friends and I were earning part-time was enough proof that it worked, but they still had to do their due diligence. I always encourage people to get the facts, but these two needed not only facts, and data to justify the facts, but also numbers to justify the data to justify the facts, and proven theories to justify the numbers to justify the data to justify the facts.

As lunch was coming to a close, I received a text message from Michael. The text simply read, "I'm ready." I smiled and laughed out loud. All three looked at me curiously. I showed them the text.

"You see this," I said. "This came from a 21-year-old friend of mine named Michael. He's a volunteer firefighter and dispatcher. He's ready to join the team."

They smirked as if to say, "A 21-year-old dispatcher. That's cute." After all, they were highly educated engineers. They were critical thinkers. This kid was clearly not on their level.

In a nonconfrontational way, I said, "You watch. While you guys spend the next month analyzing things, this kid's going to tear it up."

About one month later, one of the engineers called to ask more questions. He told me how they had been looking into things and thought this might work. However, he had a little more research to do before giving me his response.

During the conversation, he asked, "By the way, whatever happened with that kid who texted you?"

"Michael became the first 21-year-old in the history of the company to hit a nationally recognized leadership level," I answered, and then I shared how much money he had earned to date. The engineer was silent. "While you have been asking people's opinions on whether or not they would want the product you would be selling, Michael has been using the product and sharing how it has improved his life," I explained. Then I added, "You would do very well with this *if* you would just get out of your own way."

He agreed.

The engineer and his friend were only doing what they thought was right. The moral of the story is that when your desire is great enough, the facts are not as important. Michael knew he would encounter adversity. He knew he would have to learn how to make sales and overcome negativity from others. He also knew that every success story has three components: the dream, the struggle, and the prize. He did not worry about the struggle because he knew he had fortitude. The advantage always goes to the person who is not afraid to enter the arena. Michael knew what it was like to crawl deep into a building that was on fire and press forward until the job was done. He certainly was not going to be afraid of someone saying no to the product he was trying to sell.

Michael was one of the key members of our team. He became more than a friend to me; he became my brother. That was partly due to the fact that we built a large and productive sales team together, but mostly because he and I shared the same philosophy of, "It's not over until we win." This is the same mindset we have instilled in our team members.

When you say you want to do something, others will try to tell you it is not possible. Either they are going to be right, or you are. Make it you!

By the way, I am not denigrating engineers. They have brilliant minds, but they would also be the first to tell you that they overanalyze things. The exception might be my friend Les, who was a sanitation engineer. He often simplified his profession by saying there were only two things sanitation engineers have to remember in order to be successful: 1) Make sure things flow downhill, and 2) Never lick your fingers.

You may want to take a moment to get that thought out of your head before you continue reading.

Every success story has three components: the dream, the struggle, and the prize.

Commitment

According to an article by Richard Halloran in the *New York Times* on Jan. 21, 1981, Martin A. Treptow was a "young and obscure American private killed in France in World War I. . . .Treptow was born January 19, 1894, in Chippewa Falls, Wisconsin, near Bloomer, where he grew up. He was working as a barber in Cherokee, Iowa, when the United States entered the war in 1917. There he enlisted in the Iowa National Guard, which became the 168th Infantry Regiment, 84th Brigade, in the 42d 'Rainbow' Division when it was called to Federal service." On July 29, 1918, Pvt. Treptow gave his life while carrying a message between battalions under heavy artillery.

Had it not been for President Ronald Reagan's inaugural address, history might have forgotten Pvt. Treptow. According to the American Presidency Project, during his address, President Reagan said, "Those who say that we're in a time when there are no heroes, they just don't know where to look." He spoke of monuments to heroism as he mentioned "the sloping hills of Arlington National Cemetery with its row upon row of simple white markers."

When Treptow's body was found after the battle, he was carrying a diary. President Reagan explained that "on a flyleaf under the heading 'My Pledge,' he had written these words:

'America must win this war. Therefore I will work, I will save, I will sacrifice, I will endure, I will fight cheerfully and do my utmost, as if the whole issue of the struggle depended on me alone.'"

President Reagan stated, "No weapon in the arsenals of the world is so formidable as the will and moral courage of free men and women."[1] Private Treptow knew the meaning of the words *courage* and *commitment*. They became part of his legacy.

Committed people follow through on what they say they are going to do long after the mood they were in when they said it has passed. You will never develop or lead a committed team if they feel you change your mind every time you encounter resistance. It is okay to adjust your sails when the wind changes, but it is not okay to throw your hands in the air and give up just because the road to success was more difficult than you had anticipated. The world is full of people who have given up too soon, and many of them may have been on the brink of success. Unfortunately, we will never know.

Committed people follow through on what they say they are going to do long after the mood they were in when they said it has passed.

Service to Many Leads to Greatness

Why do you do what you do?

Harvey Firestone was an American businessman, and the founder of the Firestone Tire and Rubber Company. Firestone once said, "The sole reason a business exists is because it meets a human need." The success of your team will largely depend on your ability to find a human need and fulfill that need.

So let me ask you again—why do you do what you do? What need are you trying to fill? What difference are you trying to make? The service we perform as firefighters is the epitome of fulfilling human needs. Firefighters rarely become firefighters for financial prosperity; however, becoming a firefighter offers many things that money cannot buy. It is not about what you can get; it is about what you can give and who you can become. People who express an interest in firefighting or who have a child interested in firefighting often ask for my advice about the fire service.

Do people ask you for advice about your chosen career? If so, what do you tell them?

In my case, it might be easy for me to talk about the dangers of the job. In fact, it would be wrong not to. However, I always emphasize what I love about being a firefighter. I hope you and your team members also talk about why you love doing what you do. I once opened a fortune cookie that read, "Service to many leads to greatness." I really connected with that statement because in our industry, that is what we do—serve many. That is why firefighters are respected throughout the world. When you fulfill a necessary human need, like the need to feel safe, people respect you. When people respect you, it provides you with a sense of pride and accomplishment. That is just one of the reasons why those of us lucky enough to be firefighters consider it the greatest job on earth.

I encourage you and your team members to spend time connecting with the reason for your existence. Talk among yourselves about why people need your organization. Once you can answer that question, write down the reasons why you and your team love what you do and post it where everyone can see. Some might call this a mission statement, but it is more like an *existence* statement. When done correctly, this exercise can empower you and your team members.

Here are the top four reasons why I love being a firefighter. They are the reasons why I would endorse this as a career choice for others, and these are the things I mention when someone is contemplating a career in the fire service.

Top four reasons to be a firefighter

1. The people you will meet and the lives you will touch. When your job description is "protect the weak," you know you will be impacting lives. The truth is, the people we have sworn to protect are not weak. They are just having weak moments. We meet people at their worst moments, and we are there to do everything in our power to make things better. Combine those interactions with the other times a firefighter connects with members of the public, and you can begin to realize the level of impact I am talking about. The senior citizen who needs help hanging a smoke detector, the young parents who detect an unusual odor of gas in their home, the child who is injured playing in the school yard at recess—all are examples of people we serve and protect on a typical workday. And it is not just the customers who make this a great profession; it is also the other firefighters from around the world. Firefighters are pillars in their communities. They are dedicated to serving others and putting their personal safety aside for the love of their community (fig. 5–5). Most of my best friends are firefighters, and not just because it is the circle I run in. It is mainly because they are some of the most genuine, caring, and committed people I have ever known.

2. Utilizing your talents, skills, and abilities. Your value in most organizations is to be a function, not to be yourself. This profession will take you and all of your imperfections. Everyone has a unique skill set. The fire service needs you and your talents. Even if an individual has not yet identified his or her talents, but still brings a desire to contribute and a willingness to learn, that person is an asset. Those two qualities alone will give a person the foundation needed to learn new skills and gain the proficiency required of those who wear the Maltese cross. Some people come into the fire service with backgrounds in building construction. Others have never hung up a picture. It doesn't matter. The fact is the fire service has the ability to help someone develop and identify what he or she is good at in a way that few other professions can.

Fig. 5–5. Firefighters are dedicated to serving others and putting their personal safety aside for the love of their community.

3. The person you get to become in the process. Only the servicemen and servicewomen of our armed forces can understand this the way firefighters do. We exist to serve our fellow men and women at the highest level. For those in the fire service, this goes beyond what we do when we answer the call. Being a firefighter is a 24/7 calling. There are so many ways firefighters volunteer their time off duty to contribute to society. Serving seniors at a nursing home, assisting a community after a natural disaster, helping raise money for a good cause, or starting a program that promotes education are just a few examples. This is not just a job; it is a career and a calling. As difficult as this profession is, it is extremely rewarding. You are serving a purpose like no other, and you will be adding value to everything around you. This is the type of calling that will enable individuals to grow and become exceptional human beings.

4. The ultimate team experience. Firefighters risk their lives together, which creates a bond that is so tight it cannot be put into words. Firefighters do not just rely on each other to do their part in saving the lives of strangers; they rely on each other to save the lives of the people standing next to them—their team members. This transcends anything we have seen in professional sports. It is not about winning a championship. It is about completing a mission and getting home safely. Yes, we joke with each other, and if the firefighters you work with find out what bothers you, they will probably exploit that weakness in a way that would make the writers of *Saturday Night Live* laugh out loud. They do this because they are connected on a family level. They are brothers and sisters who eat, sleep, train, fight, laugh, cry, overcome, and sacrifice together. Firefighters know the meaning of the word *team* better than most people ever will.

Firefighters know the meaning of the word team better than most people ever will.

Those are things I tell individuals who ask me for information about what it is like to be a firefighter. Of course, that is just scratching the surface. My hope is that you create your own list of why you love your profession.

I once wrote a blog titled "The Greatest Job on Earth," where I listed the four reasons above and asked people to add their own reasons why they loved being a firefighter. An officer from the Chicopee Fire Department in Massachusetts wrote, "There is no greater feeling in the world than when a person in the midst of personal tragedy and sorrow gives you a heartfelt thank you. They're not thanking us because we made everything better. They're thanking us because we cared enough to try. It's extremely humbling to see the sincerity in their eyes and hear the candor in their words. There's not enough money in the world to buy the emotions that simple gesture evokes."

I could not have said it better. Don't you want people on your team to feel that type of passion about what they do?

Your Needs Come Last

One of the challenges of a career firefighter is that we often work on nights, holidays, weekends, and during important family events like birthdays and anniversaries. Taking a vacation day so you can be off for New Year's Eve is not always an option, but that is part of what you signed up for. I have worked every holiday more than once during my career. I have even worked during the Super Bowl five years in a row. Imagine the pain!

On one Super Bowl Sunday, the members of our crew had prepared a magnificent spread of food. We were excited about the game. The national anthem had just come to an end, and the players were approaching the middle of the field for the coin toss. Before the coin hit the air, our dispatcher came over the intercom to inform us of an activated fire alarm in a mixed occupancy.

Accompanied by noticeable sighs, I headed to the scene with two engine companies and one ladder company. It was a three-story mixed-occupancy apartment complex with a restaurant on the first floor. The restaurant was open, but there were only a handful of people inside, along with the cooking and wait staff. To say the sounding alarm was noticeable would be an understatement. They were actually the loudest fire alarms I had ever heard. The system was wired with battery backups, so the alarm was activated throughout the building—the restaurant, basement, and common stairwell.

We searched all the floors and found no cause for activation. This was clearly a fault in the system. We unsuccessfully tried to shut down the system from the breaker panel, but it just kept going. The owner arrived, but he did not know how to shut down or silence the alarm either. To make matters worse, many of the residents in the apartments above were home. They were, of course, complaining about the sound and wondering why we could not do anything about it.

We began the process of taking down one detector at a time to see if the alarm would stop activating. This process is similar to trying to determine which bulb on your strand of Christmas lights caused the entire strip to fail. You pull one out, and if the system does not begin to operate normally, you put it back in and rule that one out as the faulty light. We completed this same process with every detector, about 30 of them, without any success. Some of the detectors were located in hard-to-access areas. We needed to bring ladders in through windows and angle them awkwardly in narrow stairwells in order to reach the units. This made the process significantly more labor intensive than usual.

More than 45 minutes into the incident (and the Super Bowl), we were still on the scene and no closer to finding the fault or silencing the alarm. I admit, I was contemplating waving the white flag and telling the owner he would need to call for an electrician. After all, there was clearly no fire. The idea of finding an electrician to respond on Super Bowl Sunday was highly unlikely, but there was not much more we could do. We began to put the ladders back on the apparatus. Before I could share the bad news with the owner, one of my officers walked up and said, "Chief, why don't we try removing all the detectors together and putting them up one by one?"

We had never done this before and I was not sure if it would work, but I had always preached that we should not leave an incident without finding a cause. Although I had been considering sending my crews back to quarters, it took one person to step up and do the right thing. I was thankful to have him on the team because he understood our mission. I called the officers together and shared the plan with them. Not one person complained, moaned, or showed any sign of frustration. Instead, they said, "Let's do it."

We took the ladders back off the apparatus and removed every detector again. Halfway through the process of putting them back up, the alarms began sounding, and we found the faulty unit. It took more than an hour, but the problem was solved. The owner, the residents, and the few people who were there having dinner truly appreciated the dedication of our team. We returned to the station and enjoyed the second half of the Super Bowl, complaining about our lousy box pool numbers.

The moral of the story is simple. Put the needs of your customers ahead of your own, be thorough, and exceed expectations. We exist to solve problems. It would be a shame to forget the reason for your team's existence and begin to look at routine, mundane, and poorly timed incidents as an inconvenience.

Put the needs of your customers ahead of your own.

A Negative Attitude Is Poisonous

Even when you have the right attitude about your team, your mission, and yourself, negativity always seems to find a way to creep in. If micromanagement and indecision are two of the biggest killers of team initiative, negativity is a close third (fig. 5–6). As a rule of thumb, remember this: negativity can go up the chain of command but should never go down. If you have something negative to say, talk with whoever is at your level of leadership and above, but do not let that negativity spread down through your team.

You may work for terrible boss. Perhaps your boss never praises, compliments, or thanks those who work under him or her. Making matters worse, this boss might also never miss an opportunity to criticize, condemn, or complain about those same people. He or she may be famous for saying things like, "Next time you need to do it this way," and then proceeding to give you poor advice. This is the type of person who will bring out only the minimum effort and performance from subordinates. Team members will do nothing more than the bare minimum, not because they are afraid of failure, but because they are tired of shouldering the blame.

Fig. 5–6. Negativity is one of the biggest killers of team initiative.

One method of dealing with poor managers like the one just described is to listen to the criticism and then go out and meet those who work under your command with a smile. The worst thing you could do is take another person's negative attitude and let it flow down through you and infect your entire team. In essence, try to become like a water filter.

As a team leader, you will be challenged when members of your team are so negative they become toxic. The moment they walk into a room, it is like a storm cloud just hid the sun. Conversations stop, people turn away, and everyone makes excuses to avoid working on projects with them. Negative teammates carry bad attitudes that spread like a virus and devastate a team's productivity in record time. You have two choices when dealing with these types of people. You can try to cure them, or you can cut them loose.

It is not that simple when the negative person is your boss. Do not put negative, condescending individuals who happen to be in leadership positions in the same category as you would real leaders. Strong leaders are not interested in blaming and criticizing; they are busy inspiring others and leading their team through difficult times. Remember that as your own influence and areas of responsibility grow. The growth of power in your career should also come with the responsibility to help others grow.

Right Is Right

I received a call from the chief of a large fire department in the northeastern United States who was interested in having me come speak to his members about leadership. He said he had heard about me and my books and that I came highly recommended. As we spoke, I could tell he was trying to determine whether or not he was going to bring me in as a speaker.

"What question can I answer for you?" I asked, helping him open up about whatever concerns or reservations he was having.

"Well, to be candid, I haven't read any of your books, but you seem to be a very active speaker. I'd like to know who made you the authority on what is right or wrong," he asked.

It was an honest question that deserved an honest answer, so I gave him one. "Well, I don't have the right to impose my values on anybody," I said. "But the values I speak and teach about are trust, honesty, loyalty, dedication, integrity, service, and personal responsibility. You and I have never met, but if those values do not align with yours, then I'm not the right man for the job." Two weeks later, I was at his department speaking to his team about leadership.

CHAPTER 5

Building Your Legacy

You see, right is right—period. There absolutely is right and wrong in this world. Even an atheist would have a hard time arguing the fact that if you live your life following the 10 Commandments, you will be a better person than someone who habitually breaks them. Even when it comes to team building, there is absolute right and wrong. Herb Brooks, the coach of the legendary 1980 US Olympic hockey team, understood the difference. He said, "I look for players whose team name on the front of the sweater is more important than the one on the back." Brooks did not want the *best* players. He wanted the *right* players. If you are not familiar with the story, watch the movie *Miracle* (2004). You will not regret it.

There absolutely is right and wrong in this world.

It is ironic that the movie *Miracle* was released that year, because the 2004 US Olympic basketball "dream team" could have learned a lot from Coach Brooks. After the 1988 Olympics, the International Olympic Committee revised the eligibility rule to allow professional athletes to compete. In 1992, the United States fielded its first of many dream teams comprised of mostly NBA All-Stars. The 1992 team that competed in Barcelona included many elite athletes, and most were classy individuals with a team-first mentality, such as Michael Jordan, Larry Bird, John Stockton, Magic Johnson, and Karl Malone. Watching this team perform was a sight to see, and they all appeared to be sincerely happy to be part of this historic gold-medal team.

Twelve years later, in 2004, the United States fielded another basketball team consisting of the biggest names in the sport at the time. This so-called dream team, however, embarrassed the country by displaying poor attitudes, lack of desire, and selfishness. The players and coaches on this team had contracts worth a combined $680 million, and their inability to put the team first resulted in three tournament losses.

The media was relentless in discussing this team, and for good reason. In their arrogance, they disrespected the very game that gave them fame and fortune. Sadly enough, it was not because they lost but because they acted as if they did not care to even be there. It appeared that the Olympics were nothing more than an inconvenience to their jet-setting lifestyles. America would have been better off putting together a team of college standouts like Herb Brooks did with the 1980 hockey team. At the very least, they would have been proud to represent their country at the Olympics in Athens that year.

What if the 2004 dream team approached the game with the same level of respect and honor as the 1992 team? What if their individual values were intact, and each possessed qualities that are desirable in a team player? The outcome would certainly have been different, even if they still lost a game or two. There is no shame in losing; the shame comes when you put on the uniform but do not care to be playing on the team in the first place.

DTRT

One of the wonderful realities of being a firefighter is that we belong to a much larger team than any of us realize. A firefighter from Hawaii can walk into a fire station in Ohio and feel welcomed. At a fire service conference one year, the question was asked, "Would a firefighter in a jam be wrong if he walked into a firehouse and said he needed a place to sleep for the night?"

A group of us discussed the meaning of the words *brotherhood* and *sisterhood* and wondered where and when the line would be crossed. After a long discussion, one fire officer said, "Any firefighter should be able to knock on the door to any fire station in the country when they are in a jam, but the sense of community should come from the men and women who work at that station who should DTRT when another firefighter (or anyone else) is in need."

DTRT stands for "do the right thing." I would love to see more industries emulate the type of bond we often see among firefighters (fig. 5–7). Why should we go through this short life always competing against each other? Let's look for win-win scenarios. Do something for someone without any thought of a return favor. You will earn people's trust and respect for setting that type of example. Make it one of your goals to become known as a person who does the right thing. I cannot think of many other qualities that trump this one when it comes to people I would want to have on my team.

For the most part, we all know what the right thing to do is. At those times when the right thing to do is not clear, a strong team has an advantage. Members of a high-functioning team can put their heads together, come up with a plan, and take action.

Fig. 5–7. More industries need to emulate the type of bond often found among firefighters.

Pushing vs. Pulling

Dwight Eisenhower was the 34th President of the United States, serving from 1953 until 1961. He had been a five-star general in the US Army during World War II and had served as Supreme Commander of the Allied Forces in Europe. He was responsible for planning and supervising the invasion of North Africa with Operation Torch in 1942 and 1943 and for the successful invasion of France and Germany in 1944 and 1945 from the Western Front.

Eisenhower was a leader who never passed up an opportunity to share his beliefs about setting an example, and he had a notable method for educating the generals under his command. Eisenhower would take a chain and stack it in a pile on a table. He would then ask the generals, "If I push that chain, which way will it go?"

Some sized up the chain and gave their opinion as to possible direction the chain would travel. However, the correct answer to Eisenhower's question was simply that it was not possible to correctly predict which way it would go.

Eisenhower would then grab one end of the chain and ask, "If I take the chain and pull it as I move in a specific direction, which way will it go?" The answer was simple—it would follow General Eisenhower in whatever direction he went.[2]

That is one of the most simple yet effective demonstrations of leadership I have ever heard. If you push your people, you simply do not know which direction they are going to go. To truly lead them, you must show them the way by your example. Teach this concept to your team members. It does not matter if an individual is leading a team of one or one thousand. Unless you are a drill instructor trying to toughen up soldiers, leading through intimidation tends to end in failure.

Do not misinterpret this message. When you are developing your team members, you will want to push them beyond what they think they may be capable of so you can help them grow and develop into problem solvers. Most people will never know what they are truly capable of if they do not have an outside force challenging them to improve. A strong leader can become that outside force, but the leader must be willing to go to the same places he or she is trying to take others. This lesson by General Eisenhower is a reminder that most people in leadership positions are like travel agents, trying to get people to a destination they have never been to themselves. Don't be a travel agent; be a tour guide.

CHAPTER 5

Building Your Legacy

You don't lead by hitting people over the head—that's assault, not leadership.

—Dwight D. Eisenhower

A Culture of Hatred

I do not know what the culture of our country will be when you read this book. If it is within a few years of the publishing date, you will absolutely understand what I am about to say. If this book is more than 50 years old and you came upon it while cleaning out the closet of your parent's house, I hope the next few paragraphs do not make any sense. If that happens to be the case, then we have taken a turn for the better in this country.

I love America and I feel blessed every single day of my life because I live here, but our country is politically divided. Of course, it has always been to a certain degree, but I cannot imagine it has ever been as bad as it was while I was writing this book, aside from the Civil War. We live in a time when most people seem to have no tolerance whatsoever for the opposing point of view. Instead of trying to understand the wants, needs, and beliefs of another, many prefer to immediately minimize, ridicule, and discredit the other person's point of view. Some seem to think that doing so is the best way to elevate their own opinions and positions.

As a writer, one of my main objectives has always been to inspire and add value to people's lives. That is not an easy thing in today's culture. In writing books like *Step Up and Lead* and *Step Up Your Teamwork*, I share my personal experiences and opinions in a practical and straightforward manner. But in doing so, I risk offending people.

Why is this? Why should a person who is trying to share a positive message be concerned about offending people? The truth is that many people are consumed with hatred for the opposing point of view. When some people watch the news, they do not want an honest, unbiased opinion. They want to hear people agree with their own point of view. As a result, they only watch left- or right-leaning media outlets, depending on their preferences. They no longer have open minds, to their detriment. If you cannot see the other person's point of view, you will struggle in your mission of leading a team in today's society.

Nelson Mandela understood this well. Mandela was the first black president of South Africa. As depicted in the movie *Invictus*, while attending a rugby match between South Africa's Springboks and England, Mandela sees that many of the black attendees in the stadium are cheering for England's team, as the mostly white Springboks represent prejudice and apartheid in their minds. Mandela says that he did the same while imprisoned on Robben Island. However, instead of changing the team's name to satisfy the requests of his constituents, Mandela openly displays his support for the team and persuades South Africa's newly developed national sports committee to support the Springboks. He meets with the team captain and suggests that the Springboks winning the World Cup would help unite the nation.

Instead of using his power to destroy something that he once hated because it was loved by those who opposed his point of view, Nelson Mandela embraced it. In real life, as in the movie, this simple but brave act helped Mandela unite a country. I believe I am in the majority when I say that Nelson Mandela lived an extraordinary life.

Disagreeing can be healthy. In today's fire service, there is a great need for us to share our opinions. Doing so will help teams engage in intense and passionate discussions about the issues and concerns that face us as a profession. We can start by agreeing that our mission is founded on cultural values like duty, honor, loyalty, respect, integrity, personal courage, and selfless sacrifice. With this in common, firefighters should respect the opinions of others and use them as a springboard for dialog, discussion, and debate. Passionate discussions are a good thing (fig. 5–8). The things we must oppose are behaviors and opinions that do not reflect our values, such as carelessness, harassment, and selfishness.

Hatred is poison. Hatred is born from misunderstanding and fear. We must all work harder to look at, and understand, the overall picture. Try to see situations from other people's perspectives. That does not mean you have to compromise your beliefs. It simply means you will have respect for those who are working alongside you as well as the public you serve.

When you show others that you have an open mind and respect their points of view, you will be surprised to discover that they will begin to do the same. On the other hand, if you discard another person's thoughts and beliefs, that person will do the same with yours, even if you are in charge. This mistake will cause unnecessary adversity and create a culture of hatred on your team.

Fig. 5–8. Embrace the opinions of others. Doing so will help you have productive conversations that can help bring about positive change. *Courtesy:* P.J. Norwood.

For to be free is not merely to cast off one's chains, but to live in a way that respects and enhances the freedom of others.

—*Nelson Mandela*

The "I" in Team

There is a saying I am sure you are familiar with. The saying is, "There is no I in team." In the context that sentence is used, I agree with it. However, individuals need to understand that there is also a high level of personal responsibility that is necessary if you intend to enhance your teamwork.

You have read about the importance of accountability, communication, and getting involved, but the "I" in team is about you taking personal ownership of yourself as an effective team member who is an active contributor, culture creator, and problem solver.

Danny was a young firefighter who was recently sworn in. His father was a retired firefighter, and he was thrilled to be able to continue the family tradition. Although eager to learn, Danny had struggled with academics throughout most of his life. Never once did he think firefighting was going to be easy, but he did not expect it to be as intellectually challenging as it was. Fire science, fire engineering, coordinating strategy and tactics, building construction, fire behavior, flow paths, pump pressures, fire protection systems, medical treatment techniques, and conflict resolution methods all came at him so fast that he began to shut down. He was overwhelmed.

Jamie was sworn in on the same night as Danny, but her test scores topped the rest of those in the academy. Her progress was swift, and her ability to retain and recall information that she only heard once was impressive. After three months on the job, Danny and Jamie received their first evaluations. Both were as expected. Word got out that Danny was not doing well. Instead of helping him improve, the other members assigned to his shift began to make jokes about him. They gave him the nickname "Dimwit Danny." He became the target of their jokes, and since he did not want to show the others that it bothered him, he laughed along with them.

Although Jamie was not assigned to the same group or shift as Danny, she was bothered by the treatment he was receiving. She offered to help Danny, and so they met in between their shifts during the day. Jamie worked to help Danny understand and retain the information he was struggling with, and because of the extra help, Danny made rapid progress. His improvement was noted on his six-month evaluation, and his officers began to look at him as one of the go-to guys on the group when they wanted to get something done. This made Danny and his father very proud.

That story illustrates what the "I" in team is. Jamie did not have to help Danny. They were on different shifts and rarely saw each other. However, because she was a team player, she dedicated some of her personal downtime to help strengthen the overall team. Jamie put the "I" in team. How can you do the same? It's simple. Do more than you and others expect of you. Do something to benefit the organization you belong to, even if it does not appear to benefit you. That may be what it looks like on the surface, but when you do something to advance the team, you are leading by example. Do this and others will follow your lead. If they don't, eventually the cream will rise to the top, new players will emerge, your influence will grow, and your team will become stronger.

The "I" in team is about you taking personal ownership of yourself as an effective team member who is an active contributor, culture creator, and problem solver.

Be a Culture Creator

A *culture creator* is a term used to describe someone who improves the quality of life in a society or workplace. These people are great at creating environments where team members can thrive. Culture creators often are natural leaders that other people are drawn to. They do not plan, they just act, and without effort, their actions always seem to be on target. This, however, is not always the best way. As with any other form of leadership, a true culture creator must take action with specific intent.

You do not have to be the head of an organization to fulfill this role. You just have to be willing to step up when the time is right. An individual who has the ability to encourage and inspire others to take positive actions is a valuable asset to any team. You can be that person. It may be challenging at first, but when others start join in, you will begin to create momentum, and with team momentum, you can truly step up your teamwork.

Since the release of *Step Up and Lead*, I have had the privilege of speaking in front of hundreds of organizations and thousands of people. This opportunity provided me with the unique advantage of being able to spend time with effective and ineffective organizations both in and out of the fire service. In my experience, one of the defining characteristics of the more effective organizations is that they have positive, energetic people who are committed to the pursuit of excellence. They are people who lead by example and understand that it takes deliberate actions in order to achieve team success.

One of the best experiences I have had with a team of culture creators was when I visited Tampa, Florida. I had the opportunity to spend time with Captains Tony Perez, David Dittman, Bill Townsend, Jason Dougherty, and retired Tampa Fire Department (TFD) Chief Dennis Jones. These five individuals started a nonprofit organization called Friends of Tampa Firefighters. They did this with the intent of enhancing and improving training in the Tampa Bay area while uniting other departments from across the state of Florida. Based on my experience, they accomplished their goal. All of the members I met from the TFD, as well as surrounding departments, were committed, driven, passionate, and eager to learn. They were also compassionate. They welcomed me with

CHAPTER 5 Building Your Legacy

open arms and immediately made me feel like I was a part of their family. And yes, they truly were a family. Within two days I was convinced that this group had some of the strongest culture creators in our industry. Several pictures in this book feature members of the TFD, and I hope Tampa's community leaders know how lucky they are to have such an excellent fire service team.

Sometimes it is the simplest little things that will help you create the right environment for your team to thrive in. At Station 5 in Tampa, they put photos of past and present firefighters on the walls (fig. 5–9). At the Massachusetts State Fire Academy, each graduating class designs its own unique plaque to put on the wall (fig. 5–10). In Orange, New Jersey, a firefighter designed and built a table with their logo on it in an effort to improve morale (fig. 5–11). Throughout the country, many organizations issue challenge coins to firefighters as tokens of appreciation for a job well done (fig. 5–12).

Fig. 5–9. The photos of past and present firefighters are displayed on the wall at Tampa Fire Department, Station 5.

Fig. 5–10. Here are some of the plaques designed by graduating classes of the Massachusetts Firefighting Academy.

Fig. 5–11. Orange, NJ Fire Department Captain Elvin Padilla displays a table he designed for his firehouse.

Fig. 5–12. Issuing challenge coins is a great way to boost morale.

These are all examples of ways you can create the culture and environment you want. People want to have fun. They want to feel appreciated. They want to be in an environment where they are pushed to achieve more but are not ridiculed when they struggle. Be a person who creates this type of environment.

Culture creators live by a simple philosophy. They want things to improve, and they do not feel like waiting around for someone else to do it. Life is too short. They understand that if something does not exist, they simply have to build it themselves. If you are a culture creator, I encourage you to keep doing what you are doing.

If you have culture creators on your team, my advice is to identify and support them. They will do more for your organization than you can possibly imagine. They are on the front lines. They interact with and influence your team members as well as the public. Department heads should embrace these individuals and work with them. Bringing in an occasional cake for the team is great, but that type of appreciation will wear off. When you have self-motivated people creating a winning environment for all, let them run in their lanes. Do not hold them back or decrease their motivation by criticizing or micromanaging their every move. Encourage them to keep doing what they are doing.

The Step-Up Challenge

Where do you go from here? You have obviously purchased this book with the intent of learning, and I commend you for following through on that intention. Following through on what you say you are going to do is a key to success in any area of life.

If you are like most people, while reading this book, you were subconsciously putting its methods and suggestions into one of three categories: things you must keep doing, things you must stop doing, and things you must start doing.

Throughout your life, you have experienced things that work and things that do not work. You have also witnessed people doing things a different way and getting better results in the process. This is precisely why you should start each day by reviewing your goals and asking yourself these three questions:

1. What must I keep doing?

2. What must I stop doing?

3. What must I start doing?

These three questions help you take action with specific intent. You would then be wise to take it one step further and include your team in this process by participating in the culture development exercise described in chapter 1.

Next, I personally want to invite and encourage you to accept the Step-Up Challenge. This is a simple three-part challenge that will help you commit to your personal goals and those of your team. If you accept this challenge and follow through, you can transform your life. The three parts are as follows: 1) Read daily; 2) Respect and encourage others; and 3) Solve a problem.

Following through on what you say you are going to do is a key to success in any area of life.

Three parts of the challenge

1. Read daily. When billionaire Warren Buffett was asked what he did to get smarter and succeed in business, he replied, "I sit in my office and read all day." Buffett, chairman and CEO of Berkshire Hathaway Inc., is widely considered the most successful investor of the 20th century. His long-time colleague and Berkshire's vice-chairman, Charlie Munger, confirmed Buffett's comments, estimating that he spent 80% of his workday reading. The billionaire's focus is on critical thinking and a constant emphasis on building knowledge to support opinions. If one of the wealthiest investors in the world places that type of emphasis on reading, why wouldn't you?

When most people take on a new job, they read books and documents relating to their field so they can build some level of foundational knowledge. After their first few months on the job, some continue, while others stop. After the first year, most stop altogether. We all have busy lives, many responsibilities, and different priorities, but it is tragic to think that there are people in our industry who do not spend a few minutes a day reading credible fire service publications. The risks are too great and do not favor the uneducated.

What if you spent a minimum of 15 minutes a day reading? In *Step Up and Lead*, I discuss the importance of self-education and challenge readers to spend a minimum of 15 minutes a day reading the right type of book. If you do the math, 15 minutes a day × 365 days a year = 5,475 minutes a year. Imagine how much more knowledge you would have with an additional 91 hours or more of education. Now, imagine your team doing the same. If you have a team of four, that translates to more than 20,000 minutes, or more than 330 hours, of self-education that you can draw from when needed. If you think that is impressive, consider what would happen if you increased this to 30 minutes or an hour a day.

What should you read? That depends on your goals and needs. When you encounter something new that you do not know the answer to or do not understand, find the answer. One of the best ways to find that answer is in written form. Once you learn something new, pass it on to others on your team. If you did not know it, chances are others will not know it either.

2. Respect and encourage others. Your team exists to serve a purpose, and that purpose is to serve others. This part of the challenge, however, is about more than simply providing great customer service. It is about becoming the type of person others want to be around. It is about respecting everyone you come in contact with and encouraging those around you. Be a builder of people. When you make others feel good about themselves, they will want to spend more time around you. Set the right example and create a positive environment wherever you go. It is not easy to do this all the time; if it were, it would not be much of a challenge.

Value your relationships. Do not hold grudges. Be bigger than that and do the right thing. Everyone wants to be respected, yet the majority of people in this world feel underappreciated. Imagine what would happen in your world if you filled this void for those around you. Imagine how much more productive your team members would be if they were surrounded by encouragers.

Life is meant to be an enjoyable experience. Not just for you, but for those around you. People want to be around people they know, like, and trust. They want to be led by someone who cares about them. This part of the challenge can be summed up with the following sentence: Treat others the way you want to be treated.

3. Solve a problem. What are the problems that are plaguing you and your team? What challenges are you currently facing? Maybe everything is fine, but things can always be better. Take charge and fix something that is broken. From this moment on, I challenge you to no longer look at problems as anything other than opportunities to grow, change, and become better. When you begin looking for solutions instead of obstacles, it is amazing how much more you can accomplish.

The problem can be anything. Maybe you want to increase morale on your team. Perhaps you have been procrastinating and want to start taking action and right a wrong. Maybe you want to be the catalyst for creating momentum on your team. Perhaps you want to continue on your journey of personal development by picking up another book and increasing your knowledge. Problems are not difficult to find, especially in the fire service (fig. 5–13). Whatever it is, I challenge you to stop waiting for things to be perfect and take action. When others see you as a problem solver, they will respect you. When it is time to solve bigger problems, they will be more inclined to follow you. Remember that leading is actually a very simple thing that can be summed up in four steps:

1. Identify a problem.
2. Assemble a team of people.
3. Develop a solution together.
4. Solve the problem.

Leading is not complicated. The act itself can be expressed by words such as listen, choose, act, and inspire.

Fig. 5–13. The reason firefighters exist is to solve problems.

Although this is a personal challenge from me to you, it is meant to serve a bigger purpose. When you start doing these things—educating yourself, encouraging others, and solving problems—you are setting a great example for the rest of the people on your team to follow. When you accept this challenge, I want you to make yourself accountable to someone. It can be a family member, a teammate, or a friend. Verbalize your commitment. Powerful things happen when you go public with your commitments. They become real. If you are not sure who to share your enthusiasm with, feel free to reach out to me on social media. Your message can be as simple as, "I accept the Step-Up Challenge." If you want to elaborate on what goals you are setting for yourself, that is great. If not, the message is enough.

I can tell you from firsthand experience that the moment you formalize your commitment, either verbally or in writing, you will feel empowered. You will also feel obligated to follow through. Sometimes, the only missing ingredient needed in order to achieve true success is having the courage to make your goals known. Don't wait any longer, begin today. Set the example for others to learn from. Many will step up and challenge themselves in return, and that is when you will have the best chance of achieving true team success (fig. 5–14).

Fig. 5–14. Members of Kearny Fire Department, Group C, at the 2013 KFD Awards Ceremony, where they were awarded a unit citation for teamwork displayed while rescuing a construction worker on a barge the previous winter. Bottom (left to right): Firefighters Ian Kaneshige, Arthur Bloomer, William Solano, Michael Kartanowicz, James Kroll, Darell Szezypta, and Anthony Calabro. Top (left to right): DC Frank Viscuso, Capt. Drew Mullins, Chief Steve Dyl, Capt. Kevin Donnelly, and Firefighters Michael Richardson, Christopher Stopero, and Michael Kaywork. Not in photo: Captains David Kealy, Thomas McDermott, and George Harris; Firefighters William Huhn, Eugene Richard, and Richard Lowinger.

References

1. Ronald Reagan, First Inaugural Address (January 20, 1982), http://www.presidency.ucsb.edu/ws/?pid=43130.

2. Bob Davids, "Leadership without Ego Is the Rarest Commodity," TEDxESCP video (April 10, 2010), http://tedxtalks.ted.com/video/TEDxESCP-Bob-Davids-The-rarest.

INDEX

A

abandoned buildings, skill drills in, 44
absorption, of learning, 77
accommodation, 120–121, 193
accountability
    defined, 213
    freedom of speech and, 216
    individual, 213–214
    interlocking, 219
    off-duty, 214, 215–216
    rules and, 214–215
    snowstorm incident demonstrating, 207–212
    soft environment type of, 214–215
    systems of, 210–211, 212
    team collapse without, 181, 186, 188
accountability board, 210–211, 212, 213
accountability partners, 216–219
action, decisive, 131
active listening, 150
activity
    productivity compared to, 66–69
    rock demonstration regarding, 66–67
    success-producing, 68
adaptation
    failure as opportunity for, 64
    by leaders, 116, 132
    as teamwork success, 2, 21–22, 39
adversity, 226–229
affirmation, 198–200
after action review (AAR), 70–71
air bag rescue system, 9–10
Ali, Muhammad, 235
Apollo 13, 21
Armstrong, Lance, 115
athletics
    mentorship in, 106–107
    team leading in, 115
attention span, 76, 170–171
attitudes
    toward change, 126–127
    as infectious, 207, 235, 252
    during learning, 78
    of mentees, 108, 111
    negative, 11, 12, 27, 35, 37, 198–200, 250–252
authenticity, 137–138
autoignition, 3–4
Avillo, Anthony, 25, 109
avoidance
    of conflict, 190, 193, 196
    of responsibility, 181
    of stress, 83
avoiders, 205

B

backdraft, 38, 110
bad decisions, 25, 184, 232
Ball, Lucille, 231
the Beatles, 26
believability, of stories, 168
Bell, Alexander Graham, 28
Bird, Larry, 253
Bloomer, Arthur, 270
boiling liquid expanding vapor explosion (BLEVE), 188–189
boiling point, of water, 3

bonding
    among firefighters, 150–153, 247, 255
    fun and, 150–152, 265
    relationships developed by, 151–153
Boston Red Sox, 226–227
bottle rockets, 204–205
brand creation, 97–98
    customer service as, 99–100
    and marketing, 99–101
breaking it down, 81–82
Brooks, Herb, 253, 254
brotherhood, 2, 254
Brown, Les, 92, 240
Brunacini, Alan, 116, 117
Bryant, Captain, 187–188
Bryant, Paul William "Bear," 7, 64
Buffett, Warren, 267
Burg, Bob, 196
BusinessDictionary.com, 213

C

Calabro, Anthony, 270
Canfield, Jack, 162
caring, creation of, 165
Carnegie, Dale, 150, 169
Carter, Harry R., 76–77
Čavić, Milorad, 4
celebrations, 46
challenges
    coin, 263, 265
    of fire service leadership, xi
    physical, 50–52
    team-building, 50–52
    wellness, 152
change
    attitudes regarding, 126–127
    creating, 125, 127
    decisiveness during, 131
    desiring, 43, 125
    encouragement during, 131
    failure, moving beyond, and, 130
    fearing, 120, 125–128, 132
    in fire service, 2, 125
    hatred of, 125–126
    leadership directing, 116, 120, 126–127, 129–132
    momentum created by, 43
    necessity of, 125, 126–127, 128
    probationary firefighters creating, 38
    progress measured during, 131
    simplexity during, 130–131
    as stressful, 125, 128
    team players assisting, 129–130
    teamwork during, 126–127
    transparency during, 132
    vision and goals for, 130
Chief Has Arrived On Scene (CHAOS), 182
children, responsibility in, 153–154
clarity
    of goals, 39, 43
    of vision, 164–165, 181
climate, of teamwork, 219–220
coin challenges, 263, 265
collaboration, 193
collapse. *See* team collapse
collective judgment, 65–66
coma of complacency, 110
combustible courage, 235–236
commanders, 200–201
commitment, 98, 102, 103, 186, 188, 242–244, 270
communication
    coordination through, 31–33, 164
    delegation after, 164
    dialogue, discussion, and debate within, 220
    discipline required for, 33–34
    dysfunctional teams and, 191–192
    failure of, 163–164
    fire service emphasis on, 28, 31–33
    firefighter trapped due to poor, 29–31
    forms of, 163
    high tech vs. high touch in, 171–172
    ineffective, 165
    preparation prior to, 165
    storytelling as effective, 165–169
    team collapse through lack of, 163, 180, 181, 186, 191–192
    teamwork through effective, 28–33, 171–172
community relations
    demonstrations for, 52–54
    props built for, 56–58
competition
    as conflict addressing technique, 193
    healthy and unhealthy, 192
complacency, 110, 173–174, 179
compliments, 142

compound effect, 92
compromise, 191, 193, 195
compulsive pleasers, 203
confidence
    false, 231
    preparation, as creating, 87–88
    in public speaking, 169
    transparency and, 135
confidentiality, in conflict resolution, 194
conflicts
    addressing, 193–194
    avoidance of, 190, 193, 196
    competition and, 193
    confidentiality during, 194
    in customer service, 193–194
    mitigation of, 39–40, 191
    as passion, 196
    resolution of, 189–191, 193–196
    sources of, 191–192
    team collapse due to, 182, 185, 186–188, 191–192
confusion, stories as preventing, 166
control
    discipline required for, 33–34
    maintaining, 33
    relinquishing of, 95–96
    teamwork requiring, 31–33
coordination
    among companies, 17–20, 31–33, 164
    discipline required for, 33–34
coping, with problem personalities, 200–206
correction, private nature of, 163
Cortés, Hernán, 235–236
courage
    combustible, 235–236
    under fire, 233–235
    during public speaking, 234
creativity
    in public speaking, 169
    of team players, 103–104
crew rowing, 4, 237–239
cross-training, 41
culture
    development, 58, 91, 257–260, 261, 262–265
    of hatred, 257–260
curiosity, stories as creating, 167
customer service
    as brand creation, 99–100
    conflict resolution in, 193–194
    fire service and, 3, 41, 80, 99, 184, 250
    of Kearny Fire Department, 80, 99

D

daily method of operation (DMO), 90–92
de la Vega, Ralph, 228
deaths, of firefighters, xi, 214, 222–223
decisions
    bad, 25, 184, 232
    collective judgment for, 65–66
    emotion-based, 168
    headline test for, 159
    indecisiveness during, 122, 126
    ownership mentality during, 220
    as permanent, 64–66
    preparing for, 65, 184
    under pressure, 119, 184
    team involvement in, 126, 184, 220, 223, 226
    transparency during, 135
decisiveness
    of action, 131
    in leading teams, 119, 122
defensive strategy, 69, 222
delegation
    communication prior to, 164
    steps to, 212–213
    team leading through, 154
demonstrations, for community relations, 52–54
development exercise, 58, 91
dialogue, discussion, and debate, 220
differentiating yourself, 168–169
difficult people
    handling of, 196–200
    percentage of, 197
    problem personalities of, 200–206
    strategies for, 199–200
digital age, 139
dinosaurs, 35, 37, 125, 173
diplomacy, 206
disabilities, learning, 78
discipline
    communication, coordination and control requiring, 33–34
    teamwork excellence through, 139
dispatchers, 13, 14, 93–94, 161, 174
Dittman, David, 262
"do the right thing" (DTRT), 254–255

Donnelly, Kevin, 94, 270
Dougherty, Jason, 262
downtime, utilization of, 28, 47, 49–50, 89, 261
dream, compared to nightmare, 27
dream teams, 253–254
drills
    abandoned buildings for, 44
    on "close calls," 111
    by Kearny Fire Department, 49
    multiagency, 55–56
    multi-instructor, 54–55
    obstacle courses as, 51, 56, 57, 86
    real-time, 47–48
    as team goals, 48
dry lines, during fires, 174–175
Duffy, Jim, 25
Dunn, Vincent, 132–133
Dye, Gary, 110
Dyl, Steve, 270
dysfunctional teams
    boiling liquid expanding vapor explosion as, 188–189
    collapse of, 186–188
    communication problems in, 191–192
    in fire incidents, 186–188

E

Eat That Frog!: 21 Ways to Stop Procrastinating and Get More Done in Less Time (Tracy), 92
Edison, Thomas, 230
education. *See also* self-education
    coma of complacency avoided by, 110
    experience as, 40, 76, 110, 111, 112
    for leading teams, 118
    mentorship and, 110–112
    for young firefighters, 63, 110
ego, 121, 182, 192
80/20 rule, 63
Eisenhower, Dwight, 255–257
emotionality, of stories, 168
encouragement
    by accountability partners, 218
    during change, 131
    life enhancers for, 26
    for motivation, 142, 148, 161, 268
    words of, 161–163
engine companies
    coordination among, 33
    cross-training of, 41
    in layered leadership, 17–19
enthusiasm, 102
entitlement mentality, 220
entrapment. *See* extrications
equipping team members, 40
esteem needs, 144
ethics, 157
evacuation orders, 194
expectations
    criteria for, 157–158
    establishing, 154–156
    Tour Cs, 155–156
experience
    hiring based on, 74
    position title compared to, 118, 121
    stories of, 166
    as teacher, 40, 76, 110, 111, 112
extrications, 9, 48–49, 52–54, 95–96

F

failure. *See also* mistakes
    benefits of, 64
    causes of, 163
    communication lapse, and, 163–164
    learning from, 229–232
    legacy building and, 229–232
    of missions, xii
    moving beyond, change and, 130
    team collapse, communication and, 180, 181, 186, 191–192
    of teams, 179–185
    in teamwork building, 1
fairness, 157
family, firefighters as, 98–99, 156, 247, 262–263
fear
    of change, 120, 125–128, 132
    facing of, 161
    firefighters and, 233–235
    hatred born from, 259
    of public speaking, 169
    of transparency, 135
feedback, in learning, 77
Fire Department Instructors Conference (FDIC), 59–61
fire incidents
    dysfunctional teams during, 186–188
    firefighter trapped during, 29–31

Index

flashover in, 160–162
probationary firefighters in, 174–178
resources for, 15–16
as "routine," 174–179
fire service
  accountability within, 214–216
  change within, 2, 125
  communication required within, 28, 31–33
  customer service within, 3, 41, 80, 99, 184, 250
  disagreement within, 258
  generation gaps in, 192
  leadership needed in, xi, 133
  mentorship in, 106–110
  mission of, 184
  respect among, 100
  reviewing, evaluating, and revising within, 21–22
  sacrifice within, 22–23
  service in, 244–245, 246
  stress in, 51, 105, 125
  teamwork in, 1, 46–47, 247
  women in, 219–220
firefighters. *See also* probationary firefighters
  accountability partners for, 216–219
  adaptation among, 2, 21–22, 39
  advice by, 221
  bond between, 150–153, 247, 255
  brotherhood among, 2, 254
  courage of, 233–235
  dumb things said by, 221
  as family, 98–99, 156, 247, 262–263
  fear among, 233–235
  female, 219–220
  flashover event for, 160–162
  heart attacks among, 50–51, 222
  line-of-duty deaths of, xi, 214, 222–223
  motivations of, 146
  obstacle courses for, 51, 56, 57, 86
  off-duty accountability of, 214, 215–216
  off-duty service of, 247
  personal development by, 247
  as problem solvers, 269
  reasons to be, 245–248
  respect among, 100
  respect for, 216, 245
  trapped, 29–31
  uneducated, 110–111
  veteran, 35
Firestone, Harvey, 244
The Five Dysfunctions of a Team: A Leadership Fable (Lencioni), 186
flashover
  during fires, 160–162
  as full team involvement, 143
flashpoint, 142–143
flea training, 140–141, 154
followers
  of leaders, 43, 124, 136, 235, 269
  as leadership, 120, 123
force, in addressing conflict, 193
formal interviews, 188
Franklin, Benjamin, 105
freedom of speech, 216
freelancing, 73, 186, 188, 214–215
The Friend Virus (Robbins), 24–25
Friends of Tampa Firefighters, 262
frog-eating metaphor for success, 92–93
fun
  for bonding, 150–152, 265
  team collapse without, 185

G

Gandhi, Mahatma, 46, 234
generation gaps, 192
generosity, of team players, 105
goals. *See also* vision
  accountability partners for achieving, 218
  change led by, 130
  clarity of, 39, 43
  incident stabilization as, 16, 21, 184
  leading teams, through visions and, 114, 164
  for learning, 78
  life safety, 16
  momentum sustained by, 43, 44
  property conservation, 16, 21, 184
  reviewing, evaluating, and revising, 21–22
  skill drills as, 48
  teamwork development through, 5, 43–45, 47
  transparency of, 134
goldfish, attention span of, 76, 170–171
Goodyear, Scott, 4

grant writing, 42, 57, 58–59, 125
"The Greatest Job on Earth" (Viscuso), 248
Greenberg, Sidney, 118
Gross, Michael, 4
Group C, Kearny Fire Department, 155–156, 270
group recognition, 39

H

Halloran, Richard, 242
Halton, Bobby, 60
hands-on training (HOT), 60
Harris, George, 270
hatred
    of change, 125–126
    culture of, 257–260
Hayes, Bruce, 4
headline test, 159
heart attacks, among firefighters, 50–51, 222
help, calling for, 222–224
Heraclitus, 236
hierarchy of needs, 143–145
high tech vs. high touch, 171–172
Hill, Napoleon, 229
hiring, 74, 101, 124
hoselines, uncharged, 174–178
Hostetler, Jeff, 106–107
hot buttons, 143–148
How to Win Friends and Influence People (Carnegie), 150
Hudson County Peruvian Day parade, 79–80
Huhn, William, 270
humor, importance of, 105

I

I Love Lucy, 231
ideas, determining best, 95–96
identity, team collapse without, 184
improvement
    dinosaurs avoiding, 35
    momentum sustained by, 43
    team leading through, 71, 74
incident commanders (IC)
    accountability boards for, 210–211, 212, 213
    assessment considerations by, 12, 15, 20
    duties of, 13, 14–15, 16, 19, 20
    in layered leadership, 19, 20
    reviewing, evaluating, and revising by, 68–69
    self-education of, 15
incident stabilization, 16, 21, 184
incidents
    goals and objectives of, 16
    stabilization of, 16, 21, 184
inclusivity, 157–158
indecisiveness, 122, 126
individuals
    accountability of, 213–214
    motivating, 146, 148, 200
    needs of, 143–145
insubordination, 123, 207
integrity
    of team players, 101–102
    transparency as, 134
interlocking accountability, 219
International Business Machines (IBM), 230
interviews, formal, 188
It Worked for Me: In Life and Leadership (Powell), 136

J

Jersey City Fire Department, 109
Jesus Christ, 234
Johnson, Magic, 253
Jones, Dennis, 262
Jordan, Michael, 230, 253
Jung, Carl Gustav, 207

K

Kales, Stefanos N., 51
Kaneshige, Ian, 270
Kartanowicz, Mike, 19, 270
Kaywork, Michael, 270
Kealy, David, 270
Kearny Fire Department
    customer service of, 80, 99
    drills of, 49
    Group C of, 155–156, 270
    llama rescue of, 93–95
    motto of, 97–98
    props built by, 56–57
    tasks within, 19
Keller, Helen, 234
Kennedy, Robert, 125–126

key players, 118
know-it-alls, 203–204
Kroll, James, 270

L

ladder companies
  coordination among, 33
  cross-training of, 41
  in layered leadership, 20
LaSorda, Tommy, 148
Law of Probabilities, 162
lawn mowers, 26–28
layered leadership
  engine companies within, 17–19
  incident commanders in, 19, 20
  ladder companies within, 20
laziness, 88–89
leaders
  accountability of, 214
  adaptation by, 116, 132
  change directed by, 116, 120, 126–127, 129–132
  decisions under pressure by, 119, 184
  difficult people, handling of, 196–200
  downtime utilization by, 49–50
  as followers, 120, 123
  followers of, 43, 124, 136, 235, 269
  managers compared to, 17
  momentum dependent on, 43, 45
  motivation styles of, 115, 140–142
  traits of, 132
  tyrants compared to, 116
LEADERS TEACH, 132
leadership
  challenges of, xi
  change led by, 116, 120, 126–127, 129–132
  difficulties within, 3
  fire service needing, xi, 133
  improvement as integral to, 71, 74
  layered, 17–20
  mentorship programs as part of, 106–110
  readiness for, 117–118
  relinquishing control as, 95–96
  self-education for developing, 110–111, 116, 123, 267
  summed up, 116, 269
  tactical operations and, xi
  unpreparedness for, 118–123
leading teams
  ability and willingness for, 122–123
  in athletics, 115
  bonding activities when, 150–153
  by committee, 226
  compliments integral to, 142
  decisiveness when, 119, 122
  delegation when, 154, 164, 212–213
  difficult people, handling of, 196–200
  discomfort in, 120
  education necessary for, 118
  encouragement as part of, 148, 161–162
  by example, 121, 154, 256
  by expectation establishment, 154–158
  flea training compared to, 140–141, 154
  goals and visions for, 114, 164
  headline test for, 159
  hot buttons for, 143–148
  improvement as, 71, 74
  inconsistency in, 121
  key player identification when, 118
  listening as, 149–150
  people skills for, 120, 140–142, 165
  personal needs and, 120–121
  personal relationships within, 145–148
  position confused with experience in, 118, 121
  pressure tolerance of, 118–119
  priorities for, 158, 248–250
  problem solving as, 116
  public speaking as practice for, 170
  pushing vs. pulling when, 255–256
  secrecy and, 133–134
  titles and, 113, 118, 121
  traits associated with, 132
  travel agents compared to tour guides when, 123, 256
  understanding people when, 140–142
  value alignment for, 120
  vision clarity in, 164–165, 181
  words for, 160–163, 191, 195
learning. See also education
  absorption of, 77
  attention span and, 76

attitudes during, 78
  breaking it down, 81–82
  daily, importance of, 74–75, 110, 184
  disabilities, 78
  from failure, 229–232
  feedback as part of, 77
  goals and objectives for, 78
  from mistakes, 40, 74, 118, 184–185, 231, 232
  muscle memory in, 82–84
  new knowledge and, 77
  from obstacles, 86, 228
  obstacles to, 78
  order of, 77
  pace of, 76
  practice and, 76, 77
  practicing under pressure for, 85–86
  recognition/recall of, 77
  retaining information and, 76
  task fixation in, 82
  teachability and, 104
  training opportunities for, 79–80
  understanding and, 77
legacy building
  adversity as part of, 226–228
  commitment in, 242–244
  courage as, 233–235
  and culture creation, 257–260, 261, 262–265
  "do the right thing," 254–255
  failure while, 229–232
  identifying your personal, 225
  negativity as poisoning, 250–252
  prioritizing needs while, 248–250
  pushing vs. pulling in, 255–256
  reading when, 266–267
  respect and, 259, 266, 268
  rightness in, 252–254
  winning and, 240–242
legal liabilities, 71
legalities, 157
Lencioni, Patrick, 186
life enhancers, 26–28
life safety, 16, 21, 184
Lincoln, Abraham, 88, 230
line-of-duty deaths (LODD), xi, 214, 222–223
listening
  active, 150
  art of, 149

llama rescue, 93–95
Lombardi, Vince, 56
Lowinger, Richard, 270

M

Malone, Karl, 253
Management in the Fire Service (Carter and Rausch), 76–77
managers
  leaders compared to, 17
  as micromanagers, 182–183, 250, 265
  motivators compared to, 113
Mandela, Nelson, 234, 258, 260
Manley, John, 210
marketing, 99–101
Martin, Billy, 148
Maslow, Abraham, 143–145
Massachusetts Firefighting Academy, 263, 264
massive action principle, 45
Mastandrea, Joe, 110
Maxwell, John, 27, 41–42
McDermott, Tom, 110, 270
McDowell, "Big Bill," 88
McDowell, Michael, 88
mentees, choosing of, 108, 111
mentorship
  for back-up, 106
  creating opportunities for, 107
  through education, 110–112
  mentee characteristics for, 108, 111
  need for, 106, 110, 112
  programs for, 106, 107–108
  retirees and, 112
  sports-related, 106–107
  uniqueness, maintained during, 108–109
  Viscuso, F., experience of, 109–110
micromanagers, 182–183, 250, 265
Mingus, Charles, 130
Miracle (2004), 253
missions
  defining your, 113–114
  failure of, xii
  of fire service, 184
  "how" compared to "why" in, 146, 147
  incident commanders within, 13, 14–15, 16, 19, 20

Index

layered leadership within, 17–20
priority of, 12, 20
specific intent as integral to, 45, 139, 182
subteams in, 12–13
team collapse through lack of, 180
as visions, 114
mistakes, learning from, 40, 74, 118, 184–185, 231, 232
momentum
   celebrations for, 46
   change creating, 43
   competitions for, 45
   creation of, 41–42
   goals creating, 43, 44
   improvement sustaining, 43
   leaders as responsible for, 43, 45
   massive action principle for, 45
   newness for creating, 43
   performance, 42
   progress as, 45
   skill drills for, 44
   specific intent for, 45
   steps to creation of, 43–46
   strategies creating, 41
   success determined by, 41–42
   successive success for, 45–46
   training, for creation of, 42
   urgency for, 45
Monaco, Karl, 82
morale
   negative, 250–252
   team collapse and, 179–180, 186–188, 192
morals, 157
Mother Teresa, 234
motivation
   defining of, 146
   encouragement integral to, 142, 148, 161, 268
   hierarchy of needs in, 143–145
   hot buttons as, 143–148
   individuality of, 146, 148, 200
   managing compared to, 113
   of problem personalities, 200–206
   styles of, 115, 140–142
   successes for building, 45–46
Motivation and Personality (Maslow), 143–145
motor learning, 82–84
mottos, 97–98
Mulligan, Brian, 111

Mullins, Drew, 270
multiagency drills, 55–56
multiple alarms, 222–224
Munger, Charlie, 267
muscle memory, 82–84
mutual problems, in conflict resolution, 195

N

National Fire Protection Agency (NFPA), 157
National Institute of Occupational Safety and Health (NIOSH), 222–223
negativists, 198, 206
negativity
   of dinosaurs, 35, 37
   overcoming, 11, 12, 198–200
   in teams, 27, 250–252
New Jersey Devils, 55
New York Yankees, 226–227
newness, introducing, 43, 76–77
nightmare, compared to dream, 27
North Hudson Fire and Rescue, 109
Norwood, P. J., 25
nursing home fire, 208–211

O

obstacle courses, 51, 56, 57, 86
obstacles
   to learning, 78
   learning from, 86, 228
   people as, 26, 35
Obstacles Welcome: Turn Adversity to Advantage in Business and Life (de la Vega), 228
off-duty accountability, 214, 215–216
off-duty service, 247
offensiveness, stories for avoiding, 168
officers, responsibilities of, 196–197
one degree of difference, 4–5
Onieal, Denis, 163, 198
opinions, embracing of, 259
opportunities
   failure as, 64
   learning and training, 79–80
   for mentorship, 107
   problems as, 42, 226–228
Orange, New Jersey Fire Department, 264

order of information, in learning, 77
Osborn, Robert, 110
ownership mentality, 220

P

pace, of learning, 76
Padilla, Elvin, 264
Parcels, Bill, 106–107
Parks, Rosa, 234
partners, accountability, 216–219
people
    difficult, 196–200
    like-minded, 27
    as obstacles, 26, 35
    personality problems of, 200–206
    skills, 120, 140–142, 165
Peppler, Chris, 25
Perez, Tony, 262
performance
    critiques, 71
    poor, 192
    transparency for enhancing, 138
performance momentum, 42
permanent decisions, 64–66
personal connections, through stories, 168
personality differences, 192
pessimists, 206
Phelps, Michael, 4, 5
physical challenges, for health, 50–52
physiological needs, 144
piranhas, 201–202
poison, negativity as, 250–252
politics, 134, 153, 192, 257–258
post-incident analysis (PIA), 70–71
Powell, Colin, 136
power struggles, 182–183
practicing
    conflict resolution skills, 190–191
    learning by, 76, 77
    under pressure, 85–86
    props for, 56–58
    scenarios for, 49–50
    teamwork developed through, 11, 184
praise, 162–163
preparation
    80/20 rule of, 63
    communication preceded by, 165
    for confidence-building, 87–88
    decisions resulting from, 65, 184
    for public speaking, 169
    training and, 63–64
presentations. *See* public speaking
priorities
    of mission, 12, 20
    for productivity, 92–93
    setting, 158, 248–250
probationary firefighters
    as catalysts for change, 38
    education of, 63, 110
    during fire incident, 174–178
    value of, 11, 33, 35
problem personalities
    3U method of assessing, 207
    determining, 206–207
    types of, 200–206
problem solving, 268–269
    as leading teams, 116
    root cause analysis for, 72–74
    by teams, 226–228
    transparency during, 137
problems
    in communication, 165
    as opportunities, 42, 226–228
procrastination, 92–93
productivity
    activity compared to, 66–69
    prioritizing for, 92–93
    team strategies for, 68–69
progress
    measuring, during change, 131
    momentum as, 45
promotions, competitive, 45
    transparency during, 135–136
property conservation, 16, 21, 184
props, building of, 56–58
public speaking
    attention span during, 170–171
    conditions for good, 76–77
    confidence development by, 169
    courage during, 234
    fear of, 169
    practicing for, 169
    storytelling for effective, 165–169
    team leadership skills through, 170
    by Viscuso, F., 169
The Purpose-Driven Life (Warren), 102
pushing vs. pulling, 255–256

Q

The Quick and Easy Way to Effective Speaking (Carnegie), 169

R

radical transparency, 135
Ramaala, Hendrick, 4
Rausch, Erwin, 76
reading, for success, 28, 66, 120, 266–267
Reagan, Ronald, 243
recognition, group, 39
recognition/recall, in learning, 77
Reilly, Firefighter, 187–188
relationships
    bonding for, 151–153
    personal, within teams, 145–148
    transparency benefiting, 137–138
reliability, of team players, 105
religion, 153, 192
resistance
    to change, 125–128
    by dinosaurs, 35
    overcoming, 167
respect
    between firefighters, 100
    for firefighters, 216, 245
    legacy building through, 259, 266, 268
    listening as, 149
    team collapse by lack of, 179
    transparency building, 138
responsibility
    children and, 153–154
    giving of, 153–154
    team collapse by avoiding, 181
results, inattention to, 186
retirees, 112
review, evaluate, and revise (RER), 21–22, 68–69, 195
rewarding excellence, 39
Ricci, Frank, 25
Richard, Eugene, 270
Richardson, Michael, 270
Richter, Jean Paul Friedrich, 234
rightness, 252–254
Robbins, Mel, 24–25
rock demonstration, 66–67
roof ventilation skill drills, 44, 47–48
Roosevelt, Franklin D., 124
root cause analysis (RCA), 72–74
routine fires, 174–179, 221
rowing, crew, 4, 237–239
rule of five, 24–25
rules, importance of, 214–215

S

sacrifice, 22–24
sales team development, 240–242
Sanders, Col., 230
sandwich technique, 203
scenario practice, 49–50
Schwab, Charles, 148
secrecy, 133–134
security needs, 144
self-actualization needs, 145
self-convicting questions, 202
self-education
    of incident commanders, 15
    for leadership development, 110–111, 116, 123, 267
    reading as, 28, 66, 120, 266–267
    seminars for, 60, 90
    strategy development through, 184
    for success, 28
    of teams, 66, 88
seminars
    for self-education, 60, 90
    of Viscuso, F., 61, 112, 132–134
separation, for conflict resolution, 193
service, 244–245, 246
settling, 93–95
Simms, Phil, 106–107
simplicity, 130–131
sisterhood, 2, 254
size-up
    and line-of-duty deaths, 222–223
    strategies for, 49
skepticism, stories as bypassing, 167
skills
    abandoned buildings for, 44
    conflict resolution, 190–191
    continuous training of, 74, 75
    drills for building, 44
    of team players, 103
    time allotment for, 67
    unused, 78
snowstorm incident, 207–212
social needs, 144
soft environment accountability, 214–215

Solano, William, 270
specific intent, 45, 139, 182
sports. *See* athletics
stages, of team development, 34–38
standard operating procedures (SOPs), 66, 222–223
standards
    National Fire Protection Agency, 157
    for team expectations, 157–158
Step Up and Lead (Viscuso), xi, 107, 113, 132, 188, 207, 257, 262, 267
Stockton, John, 253
Stopero, Christopher, 270
storming stage, 196
storytelling, 165–169
strategies
    defensive, 69, 222
    definition of, 38–39
    for difficult people, 199–200
    incident commanders determining, 14
    momentum created by, 41
    for productivity, 68–69
    revision of, 22, 68–69
    self-education for, 184
    size-up, 49
    tactically persuasive, 34
    team collapse by improper, 184
    for teamwork, 38–41
stress
    change and, 125, 128
    decision-making as, 184
    in fire service, 51, 105, 125
    muscle memory for avoiding, 83
subordinate interviews, 188
subteams, 12–13, 81
success
    accountability for, 213–214
    achieving successive, 45–46
    after action review for, 70–71
    best idea for, 95–96
    brand creation and, 97–98
    daily method of operation for, 90–92
    in decision-making, 64–66
    education for, 110–112
    and false confidence, 231
    frog-eating metaphor for, 92–93
    laziness, preventing, 88–89
    like-minded people for building, 27
    marketing your company for, 99–101
    mentorship programs for, 106–110
    momentum determining, 41–42
    motivation built by, 45–46
    mottos for, 97–98
    productivity compared to activity for, 66–69
    reading for, 28, 66, 120, 266–267
    self-education for, 28
    settling compared to, 93–95
    succession planning for, 106–110
    team player qualities for, 101–106
    team training for, 185
Success, 24–25
succession planning, 106–107, 111
successors. *See* mentorship
Sullenberger, Chesley B. "Sully," 87
Super Bowl Sunday, working during, 248–250
Szezypta, Darell, 270

T

tactical operations, leadership and, xi
tactically persuasive strategy, 34
tactics. *See* strategies
talents, skills, and abilities (TSA), 103, 104, 246
Tampa Fire Department (TFD), 262–263
task fixation, 82
tattoos, 64–65
teachability, of team players, 104
teaching, through storytelling, 165–169
team collapse
    accountability lacking in, 181, 186, 188
    boiling liquid expanding vapor explosion as, 188–189
    causes of, 179–185
    commitment lacking in, 186
    communication failure in, 163, 180, 181, 186, 191–192
    complacency causing, 173–174, 179
    conflict resolution preventing, 189–191, 193–196
    conflicts resulting in, 182, 185, 186–188, 191–192
    delegation steps preventing, 212–213
    drama causing, 180
    dysfunctional teams and, 186–188
    freelancing as, 186

Index

fun lacking in, 185
identity lacking in, 184
micromanagers causing, 182–183
mission lacking in, 180
morale lacking in, 179–180, 186–188, 192
power struggles in, 182–183
responsibility avoidance and, 181
results, inattention to in, 186
specific intent, to avoid, 182
strategy failure in, 184
training lacking in, 184
trust lacking in, 185, 186–188
vision lacking in, 180–181
team leaders
    compromise by, 191
    conflict mitigation by, 39–40, 191
    negativity mitigation by, 11, 12, 27, 35, 37, 198–200, 250–252
team players
    change assisted by, 129–130
    choosing, 124
    commitment of, 102, 103
    creativity of, 103–104
    enthusiasm of, 102
    generosity of, 105
    humor of, 105
    integrity of, 101–102
    key player identification among, 118
    reliability of, 105
    skillfulness of, 103
    teachability of, 104
    willingness of, 105–106
team-building exercises
    collaboration and/or compromise as, 193
    community relations demonstrations as, 52–54
    conference attendance as, 59–61
    culture development by, 58, 91
    grant proposal writing as, 58–59
    multiagency drills for, 55–56
    multi-instructor drills for, 54–55
    outings as, 150–153
    physical challenges for, 50–52
    real-time drills as, 47–48
    scenario practice as, 49–50
teams
    bad, good, and great, 23
    as boiling liquid expanding vapor explosion, 188–189
    bonding activities for, 150–153
    communication as critical for, 163, 171–172
    coordination among, 33
    courage within, 235–236
    decision-making by, 126, 184, 220, 223, 226
    defining (or developing) of, 24–26
    definition of, 22–23
    difficult people within, 196–200
    dinosaurs within, 35, 37, 125, 173
    expectation criteria for, 157–158
    expectations established for, 154–156
    flashover in, 143
    focused, 47
    hot buttons in, 143, 145
    "I" in, 260–261
    individual's needs within, 143–145
    legacies of, 225
    mission failure in, xii
    negativity within, 27, 35, 37, 198, 206, 250–252
    probationary firefighters within, 11, 33, 35, 38
    problem personalities within, 200–206
    problem solving among, 226–228
    rule of five within, 24–25
    sacrifice among, 23–24
    self-education of, 66, 68
    stages of, 34–38
    synonyms for, 23
    teammate selection for, 25–26
    transparency, as strengthening, 137
    uniqueness within, 183
    veteran firefighters within, 35
teamwork
    accountability partners in, 216–219
    adaptation for successful, 2, 21–22, 39
    during change, 126–127
    communication for effective, 28–33, 171–172
    control required for, 31–33
    discipline required for excellent, 139
    failure in building, 1
    fire service and, 1, 46–47, 247
    goal development within, 5, 43–45, 47
    importance of, 1

interlocking accountability in, 219
layered leadership within, 17–20
mission first in, 12, 20
practice, for developing, 11, 184
steps to successful, 11–12
strategies for, 38–41
subteams in, 12–13, 81
10-second rule, 195–196
Tergat. Paul, 4
Terpak, Mike, 109
"A Theory of Human Motivation" (Maslow), 143–145
Thoreau, Henry David, 149
3U method, 207
time outs, 195
titles, in leadership, 113, 118, 121
Tour C, Kearny Fire Department, 155–156, 270
tour commander, 7–11, 69, 155–156
tour guides, 123, 256
Townsend, Bill, 262
Tracy, Brian, 92, 153
training
    continuous nature of, 74, 75
    laziness, prevented by, 88–89
    momentum created by, 42
    opportunities for, 79–80
    preparation as, 63–64
    teachability important for, 104
    team collapse through poor, 184
    training officer for, 74–75, 86
training officer (TO), 74–75, 86
transparency
    authenticity as, 137–138
    benefits of, 137–138
    during change, 132
    confidence during, 135
    decisions and, 135
    in digital age, 139
    fear of, 135
    of goals, 134
    integrity in, 134
    performance enhanced by, 138
    problems solved quicker by, 137
    promotions and, 135–136
    radical transparency compared to, 135
    "realness" of leaders as, 134–135
    specific intent and, 139
    teams strengthened by, 137
    trust created by, 134, 136–137, 138
travel agents, 123, 256

Treptow, Martin A., 242–243
truss roofs, 186
trust
    integrity creating, 101
    team collapse without, 185, 186–188
    transparency creating, 134, 136–137, 138
The 21 Irrefutable Laws of Leadership (Maxwell), 41–42

U

understanding
    in learning, 77
    people, 140–142
uneducated firefighters, dangers of, 110–111
uniqueness
    maintaining your, 108–109, 123
    among team members, 183
University of Charleston, West Virginia, 237–238
Unser, Al, Jr., 4
urgency development, 45

V

values
    alignment of, 120
    differences in, 192
Varè, Daniele, 206
veteran firefighters, 35
victory celebrations, 46
Viscuso, Frank
    crew rowing by, 4, 237–239
    experience of, xi–xii, 2
    mentors of, 109–110
    public speaking by, 169
    seminars by, 61, 112, 132–134
    as tour commander, 7–11, 69, 155–156
    website of, 270
Viscuso, Joe, 109
vision
    change led by, 130
    clarity of, 164–165, 181
    defined, 180–181
    goals compared with, 180–181
    team collapse by lack of, 180–181
    teamwork, goals and, 5, 114, 164

W

wall punchers, 202–203
Wallace, William, 234
Walton, Sam, 230
Warren, Rick, 102
Washington, George, 234
Watson, Thomas, Sr., 230
website, of Viscuso, F., 270
Welch, Jack, 108–109
wellness challenges, 152
whiners, 205–206
Wilcox, Frederick, 131
willingness
    for leading teams, 122–123
    of team players, 105–106
winning, 240–242
Winters, Dick, 138
women, in fire service, 219–220
Wooden, John, 22, 68, 115, 146
words, power of, 160–163, 171–172, 191, 195
wrestling club, 81–82

Y

yes men, 203

Z

Ziglar, Zig, 115, 230